I0471614

Planetary Climate before the Space Age

Ralph D. Lorenz

Copyright © 2017 Ralph D. Lorenz

All rights reserved.

ISBN: 1546814191
ISBN-13: 978-1546814191

DEDICATION

For Mum

ALSO BY RALPH LORENZ

R. D. Lorenz and J. Mitton, Lifting Titan's Veil, Cambridge University Press, May 2002, 260p.

D. M. Harland and R. D. Lorenz, Space System Failures, Praxis-Springer, April 2005, 400p. Second printing 2006 ; Chinese edition 2007.

R. D. Lorenz, Spinning Flight : Dynamics of Frisbees, Boomerangs, Samaras and Skipping Stones, Springer/Copernicus, September 2006, 346p.

A. J. Ball, J. R. C. Garry, R. D. Lorenz and V. V. Kerzhanovich, Planetary Landers and Entry Probes, Cambridge University Press, May 2007, 364p. Paperback edition 2009.

R. D. Lorenz and J. Mitton, Titan Unveiled, Princeton University Press, April 2008, 280p. Revised Paperback edition, 2010.

R. D. Lorenz and J. Zimbelman, Dune Worlds : How Wind-Blown Sand Shapes Planetary Landscapes, Praxis-Springer, May 2014, 308p., 330 illus.

R. D. Lorenz, NASA/ESA/ASI Cassini-Huygens:1997-2017 Owners' Workshop Manual (Cassini orbiter, Huygens probe and future exploration concepts), Haynes UK, April 2017. 192p. 325 illus.

CONTENTS

PREFACE

This project started as an introductory chapter to a book on the physics of planetary climate. It then grew to become a (shortly to be completed) book on the history of planetary climate, and then, in a process of amoeba-like successive division and growth, it became a book on the history of planetary climate before the space age! Although my day job is in the practical implementation of planetary exploration today, it was hugely stimulating to lay out the ingenious early efforts to understand the conditions on other worlds, and our own, before the space age began. As I researched the progressive understanding of temperature, atmospheres and radiation, I was struck by the remarkable and often daring activities of these early scientific adventurers, and the other topics on which they worked. Finding one fascinating link after another, I was reminded of James Burke's "Connections", which I loved as a teen. I hope you enjoy the journey too.

Numerous colleagues have over the years pointed me to interesting episodes in the history of planetary climate: Mark Bullock, Dave Grinspoon, Dave Catling and Tyler Robinson deserve particular mention. The text benefited from comments from Laura Kerber, Conor Nixon and Jani Radebaugh, as well as reviewers at Princeton and Cambridge University Press. Zibi Turtle's patience with the book's gestation is much appreciated.

In the event that readers wish to pursue the topic further, I have appended detailed references, sprinkled with a few additional remarks, at the end of the book.

A PLANETARY CLIMATE TIMELINE

440 BC Herodotus considers Nile flood lagging seasons, notes sea-shells on hills and salty soil indicating Egypt once undewater, recognizing terrestrial conditions vary with time

350 BC – Aristotle writes Meteorology, a work which discusses a range of phenomena in the sky, including meteors as well as weather and climate.

c. 80 AD – Chinese philosopher Wang Chong states that rain is evaporated from water on the earth into the air and forms clouds and rain

1021 – Alhazen investigates refraction of light and twilight, and deduces an effective height of the Earth's atmosphere as about 20 miles

1450 – Leone Battista Alberti developed a swinging-plate gauge, considered the first anemometer.

1604 – Cornelis Drebbel writes a Treatise on the Elements, noting the tremendous expansion of water into vapor. Drebbel went on to invent a submarine, and a self-regulating oven with a thermostat.

1607 – Galileo Galilei constructs a thermoscope, defines heat as distinct property of matter, rather than as one of Aristotle's elements (fire, water, air, and earth). Observes mountains on the moon, phases of Venus

1643 – Evangelista Torricelli invents the mercury barometer.

1648 – Blaise Pascal rediscovers that atmospheric pressure decreases with height, and deduces that there is a vacuum above the atmosphere

1650-1690s – Christiaan Huygens discovers Titan, contemplates conditions on other worlds, recognizes atmosphere and fluid may be made of different substances, different densities etc. Calculates how long a bullet would take to reach the planets

1660s – Cassini observes polar caps on Mars

1686 – Edmund Halley makes a systematic study of the trade winds and monsoons and identifies solar heating as the cause of atmospheric motions. Defines the relationship between barometric pressure and height above sea level

1724 – Gabriel Fahrenheit creates reliable scale for measuring temperature with a mercury-type thermometer.

1729 – Pierre Bouger pioneers light measurement, and develops the law of attenuation of light by atmospheric absorption (later called the Beer-Lambert law)

1735 – The first idealized explanation of global circulation was the study of the Trade winds by George Hadley.

1742 – Anders Celsius, a Swedish astronomer, proposed the Celsius temperature scale which led to the current Celsius scale.

1743 – Benjamin Franklin is prevented from seeing a lunar eclipse by a hurricane, he decides that cyclones move in a contrary manner to the winds at their periphery.

1755 – Tobias Mayer calculates dependence of temperatures on latitude and altitude

1761 – Joseph Black discovers that ice absorbs heat without changing its temperature when melting.

1761 – Lomonosov observes transit of Venus, infers presence of atmosphere

1767 – Horace-Benedict de Saussure devises a solar oven, and finds solar heat independent of air temperature

1777 – Antoine Lavoisier discovers oxygen and develops an explanation for combustion.

1780 – Charles Theodor, the Elector of Palatine, charters the first international network of meteorological observers, lasting until 1795.

1783 – Antoine Lavoisier proposes a caloric theory of heat.

1783 – Montgolfier brothers and Jacques Charles fly hot-air and light-gas balloons

1783 – Benjamin Franklin notes Laki eruption as a possible cause for cold winters

1783 – First hair hygrometer demonstrated by Horace-Bénédict de Saussure.

1785 – James Hutton describes geological cycles, argues that Earth is much older than generally thought

1787 – Thomas Jefferson maintains weather records, suggests US climate has changed due to land clearing

1787 – Richard Kirwan compiles a catalog of world temperatures

1780s – Alexander von Humboldt's explorations

1796 – William Herschel reports a catalog of stellar variability, noting that stars may be a window into variations on the sun.

1801 – Herschel notes a correlation between sunspots and the price of wheat, suggesting a direct influence of solar variability on Earth's weather

1804 – John Leslie observes that a matte black surface radiates heat more effectively than a polished surface, suggesting the importance of black body radiation

1817 – Alexander von Humboldt publishes a global map of average temperatures

1820 – Daniell develops Dew-Point Hygrometer

1822 – Fourier describes paradigm of climate as balance of heat from sun and interior versus loss to space, noting the resistance of the atmosphere to these fluxes

1824 – Sadi Carnot analyzes the efficiency of steam engines and lays the foundations of thermodynamics, describes the weather as a heat engine

1837 – Agassiz proposes ice ages, Adhemar considers astronomical change as cause

1837 – Samuel Morse develops electrical telegraph system

1847 – Ebelmen outlines the carbon cycle

1847 – Francis Ronalds and William Radcliffe Birt described a stable kite to make observations at altitude using self-recording instruments

1848 – William Thomson extends the concept of absolute zero from gases to all substances.

1849 – Smithsonian Institution begins to establish an observation network across the United States, with 150 observers via telegraph, under the leadership of Joseph Henry.

1853 – The first International Meteorological Conference was held in Brussels at the initiative of Matthew Fontaine Maury, U.S. Navy, recommending standard observing formats for weather reports

1856 – William Ferrel publishes his essay on the winds and the currents of the oceans.

1856 – Eunice Foote notes absorption of sunlight by water vapor and carbon dioxide

1856 – Piazzi Smyth suggests correlation of rock temperatures with sunspot. Makes high-altitude measurements of sunlight.

1859 – John Tyndall measures thermal absorption of carbon dioxide and other gases, the fundamental mechanism of the greenhouse effect

1864 – James Croll calculates heat flux in ocean circulation, combines geological and astronomical analyses to develop astronomical theory of ice ages, proposes ice-albedo feedback as amplification mechanism

1873 – Challenger expedition maps deep ocean temperatures

1879 – Stefan determines fourth-power law of heat radiation

1885 – Christiansen in Denmark makes first 'modern' calculation of planet temperatures using Stefan law

1880s – Langley, Rosse observe heat from the moon. Langley measures infrared spectrum.

1896 – Arrhenius calculates climate change in response to CO2 doubling

1902 – Richard Assmann and Léon Teisserenc de Bort, independently discover the stratosphere with balloons

1907 – Jose Comas-Sola detects limb-darkening on Titan, suggesting atmosphere

1900s – Lowell popularizes discussion of Mars' climate as clement. Contested by Poynting, Douglass, Alfred Russel Wallace, Comas-Sola.

1922 – Richardson pioneers numerical weather prediction

1920s – Milankovich refines calculations of astronomical forcing of climate on Earth

1923 – Pettit and Nicholson measure temperatures on Mars with 100-inch Hooker telescope

1922 – Lewis Fry Richardson organises the first numerical weather prediction experiment.

1923 – Gilbert Walker identifies a correlation ('teleconnection') between Indian monsoon and Pacific pressure readings, a feature of the El Nino/Southern Oscillation

1930 – Pavel Molchanov invents and launches the first radiosonde to a height of 7.8 kilometers measuring temperature there

1938 – Callendar proposes CO_2 levels and temperature are rising over several decades, CO_2 rise consistent with human cause, CO_2 rise is cause of temperature increase.

1944 – German V-2 rockets reach high altitudes en route to London

1940s – Development of infrared detectors for weapon aiming and tracking. Further measurement of infrared absorption in the atmosphere. Development of electronic computers.

1952 – Opik makes energy-balance model (popularized by Budyko and Sellers ~1970)

1950s – First polar ice cores and electronic numerical weather predictions. Radio measurements of Venus high temperatures

1957 – International Geophysical Year: Sputnik 1. Keeler begins systematic CO_2 measurements

1962-66 – Planetary exploration begins - Mariner 2 confirms Venus high surface temperature ; Sagan begins greenhouse calculations. Mariner 4 measures Martian surface pressure.

Planetary Climate Timeline

1

LEARNING THE EARTH, OCEANS AND SKY

440 BC to ~1780

Climate is a subject that concerns everyone, and considerations of climate problems enter the some of the first intellectual records in existence. For example, Herodotus' Histories, written (or at least told) around the 440s BC, document aspects of geography as well as history. In particular, Herodotus notes the nonintuitive behavior of the Nile river, that its level begins to rise around the summer solstice and does so for a hundred days, then the level falls throughout winter. Herodotus considers whether winds could be a factor : in fact summer snowmelt is responsible but he (incorrectly) dismisses snow, since it was known that the further south one got, the hotter and drier, so how could there be snow?

Interestingly, Herodotus also ventures some speculations on the origins and past of Egypt, noting that the soil is dark, as one might expect alluvial (water-delivered) deposits to be, in contrast to the windblown sands to the west in Libya. Furthermore he notes that shells are found on hilltops and that there is salt in the soil, and ventures that much of Egypt was once underwater. He recognizes (church dogma not yet having set an age of the Earth of only four millenia before his time) that these changes may have taken tens of thousands of years.

A more focused work, and generally acknowledged as the first serious work on the topic (~350 BC), is the Meteorologia compiled by Aristotle. This treatise covers a wide range of geophysical and astronomical topics such as the saltiness of the sea and the nature of comets as well as the nature of wind and rain (the separation of the study of meteors and meteorites into disciplines distinct from 'meteorology' came later as science became more specialized). Naturally, since the concepts of heat and energy were unknown, and the description of matter was limited to four elements (Earth, Air, Fire and Water), Aristotle's explanations are flawed, but he does a fair job of laying out some interesting problems (for example, why should hailstones, made of ice and obviously associated with cold, be more common during summer than winter?) He captures the essence of the hydrological cycle as follows:

Now the sun, moving as it does, sets up processes of change and becoming and decay, and by its agency the finest and sweetest water is every day carried up and is dissolved into vapour and rises to the upper region, where it is condensed again by the cold and so returns to the earth

After Aristotle's compilation, there was little substantial progress in the Western world. Yet, as in other fields, many Muslim philosophers and scientists made steps towards understanding the natural world, notably from the 10th through 15th centuries. The scholar Ibn al-Haytham ('Alhazen') in what is now Iraq used geometric arguments[1] to deduce the thickness of the sensible atmosphere. Knowing the diameter of the Earth, and the fact that twilight began or ended with the sun 19 degrees below the horizon, he deduced that the effective scattering height of the atmosphere could be no more than 52,000 paces. Indeed at such heights – not physically reached for almost another thousand years - the air density is only a fraction of one per cent of what it is at the ground. All in all, this was a remarkably successful exercise in reasoning.

Similarly, in China, some relevant ideas were emerging. In 1074, Shen Kuo[2] reasoned – like Herodotus - that a belt of bivalve fossil shells in mountains inland implied that the terrain there must have once been a seashore, and that mountains must be eroded and sometimes uplifted (in the West, as we see later in this chapter, Jame Hutton, seven centuries later reached similar conclusions and took them further). He also recognized that petrified bamboo, revealed by a landslide in a place where bamboo does not grow today, implied that the climate had been different in the past.

The pivotal role of the sun in controlling climate was obvious even to the ancients, but systematic consideration of the climate of Earth and of other worlds relied on correctly laying out the architecture of the solar system. While some Greek scientists got this right, it was not until the Copernican revolution beginning in 1543 that progress could really be made. An additional early challenge to understanding climate is that diurnal changes of wind, are in near-synchrony with tidal changes in the sea, and thus the role of the moon in influencing winds was initially thought to be significant.

With the Copernican layout of the Universe established with the sun at the center, backed by Kepler's laws, it became possible in the 1600s to consider the other planets in their proper place. The invention of the telescope[3] and Galileo's observation of crescent phases of Venus made it clear that Venus was closer to the sun than is Earth, whereas Mars always appeared almost completely round, implying it was further away.

The development of instruments not only augmented astronomers' vision, but also enabled the quantitative measurement of the world. Instruments to first visualize, and later measure, temperature were also first developed around this time. Galileo developed a thermoscope (only with the addition of a numerical scale does it become a thermometer), while an ingenious Dutchman working in England in the early 1600s, Cornelis Drebbel, devised an oven with a mercury tube that not only displayed a measure of heat, but regulated the air flow into it – a primitive thermostat and perhaps the first example of a feedback control system. Drebbel also ground lenses for optical instruments, showing some to his visiting countryman, the diplomat Constantyn Huygens. Remarkably, Drebbel is also reported to have invented a submarine, and conveyed King James I under the surface of the Thames in it : the depth of the submersible was indicated with a tube of mercury, in effect a barometer[4].

Evangelista Torricelli in Tuscany, confronting the problem that suction pumps could only lift water 10 meters or so, showed in 1643 that a column of mercury in a tube closed at one end was limited to only 76cm long – the same weight per unit area as 10 meters of water – and thus that this had to be the same as the weight per area, or pressure, of the atmosphere. Thus began the systematic, and eventually quantitative, understanding of the physics of atmospheres : he famously remarked that "We live submerged at the bottom of an ocean of air."

Galileo's use of the telescope in 1610, revealing that Jupiter had moons, and that our own moon is a world with mountains and craters, stimulated thinking about planets as places, and thus about their conditions. The Polish astronomer Hevelius documented the topography of the moon in detail in 1647. This widening perspective was perhaps most enthusiastically embraced by the Dutch astronomer Christiaan Huygens. Huygens ground his own mirrors, and first measured the rotation rate of the planet Mars, finding that the length of its day was rather similar to our own, and in 1655 he discovered Titan. He correctly imagined how seasons and shadow would work on other worlds. Then, in the mid-1660s, with Jean-Dominique Cassini in Paris, polar caps were detected on Mars.

Huygens considered[5] that other worlds might have seas composed of fluids with properties different from water: *"Every Planet therefore must have its Waters of such a temper, as to be proportioned to its Heat: Jupiter's and Saturn's must be of such a Nature as not to be liable to Frost...."*, a remark which rather nicely anticipates methane rain on Titan. He even speculates that there is a lot of room in between the densities of air and water on Earth for fluids on other planets – that *"The Sea perhaps may have such a fluid lying on it, which tho' ten times lighter than Water, may be a hundred Times heavier than Air'* and notes that on worlds with denser atmospheres, it would be easier for birds to fly. Indeed, this is not too far off what Titan is like today : Titan's air is 4x denser than ours, and its seas about half as dense. So Titan's seas are indeed about 100 times denser than its air, rather than the ~800 times for Earth. If birds – or even people, could breathe and not freeze, they would find that Titan's low gravity makes it a very easy place in which to fly.

Huygens closes his book with breathless happiness, *"What a wonderful and amazing Scheme have we here of the magnificent Vastness of the Universe! So many Suns, so many Earths, and every one of them stock'd with so many Herbs, Trees and Animals, and adorn'd with so many Seas and Mountains! And how must our Wonder and Admiration be increased when we consider the prodigious Distance and Multitude of the Stars?"*.

Huygens notes, in an appealingly self-consistent if slightly optimistic argument, that the astronomers on hot Mercury, so close to the sun, would think that the Earth would be inhospitably cold. And yet here we are, and therefore it follows that while we might think Jupiter or Saturn too cold to support life, it might nonetheless be present, albeit in somewhat adapted form. Huygens draws on the fact that plants and animals had been found in another new world - America - that were quite different from those known in Europe, and wonders if the differences between lifeforms on different planets might actually be less than the differences between those from widely separated parts of the Earth.

Huygens outlook, one we might call an extreme NeoCopernican, was not limited to astrobiologists on other worlds, but also sailors who probably used similar Sails and Anchors, Pullies and Rudders on other seas. He even expresses some jealousy of '*the great Advantages Jupiter and Saturn have for Sailing, in having so many Moons to direct their Course, by whose Guidance they may attain easily to the Knowledge that we are not Masters of, of the Longitude of Places*' (the problem of determining longitude at sea was not to be practically solved for another several decades[6]). Yet, while his imagination ranged across the solar system, Huygens was well-grounded enough to recognize the practical challenges of traveling to other worlds. He noted that a bullet (the fastest object with which he was familiar – that traversed a hundred fathoms in a heartbeat) would take 250 years to reach Saturn.

Huygens was not alone in his age – an English clergyman and co-founder of the Royal Society, John Wilkins, imagined the moon might be inhabited, and while recognizing[7] that the upper air was cold and thin and that there would be no Inns en route to offer victuals and shelter, affirmed '*it possible to make a Flying Chariot, in which a Man may sit, and give such a Motion unto it, as shall convey him through the Air. And this perhaps might be made large enough to carry diverse Men at the same time.....*'. He went on to make the remarkable prediction '*Supposing a Man could fly, or by any other means, raise himself twenty miles upwards, or thereabouts, it were possible for him to come unto the Moon'* - three centuries later these two landmarks in aviation were in fact met only thirteen years apart.[8]

The exploration and development of the New World gave opportunities to those who went there, as well as broadening the minds of those who stayed behind. The overwhelming impression was one of sometimes more variable weather than Europeans were used to – even just the title of one account, by James MacSparran, says it all : *'America Dissected, Being a full and true account of the American Colonies, shewing the intemperance of the climates, excessive heat and cold, and sudden violent changes of weather, terrible and mischievous thunder and lightning, bad and unwholesome air, destructive to human bodies, etc.'*

An understanding of the Earth's climate at the planetary scale first requires a realization that the Earth is round. While this was of course understood by many prior to the age of exploration, the quantitative implications in terms of the amount of sunlight deposited at different latitudes requires trigonometry, and was first calculated by astronomer Edmond Halley. In a paper published in 1693 he showed by geometric arguments[9] how the combination of the declination of the sun and the varying length of day can conspire to deliver more sunlight (averaged over a 24-hour period) to the pole at midsummer than the equator. However, the nature of heat was not yet fully understood, and so the exact implications for temperature had to wait another century or two.

Halley may have also been the first to quantitatively consider the hydrological cycle. More particularly, he adopted the thoroughly modern approach of obtaining experimental data, and applying it to a global-scale model[10]. Specifically, he warmed an eight-inch pan of water on a set of scales to the temperature of a hot summer day, regulating the temperature by holding coals nearby, and observed that in the course of two hours the pan lost half a Troy ounce of liquid.

This value of 1/10 of an inch per day is rather close to the terrestrial average flux of evaporation and precipitation, of about one meter per year. That said, the agreement is perhaps fortuitous – although Halley notes that he made the water salty like the sea, he does not record the weather conditions such as wind or humidity, which we now know would substantially affect such an experiment.

Halley then bravely upscales this measurement to the Mediterranean sea, which he briskly approximates as a 4x40 degree box (one degree being 69 English miles) to determine that the Mediterranean should lose 5280 Million tons of water per day[11]. He estimated the water influx via large rivers (scaling from the Thames, whose depth and flowspeed were known to him by personal experience, having experimented with diving bells for salvage) and concluded that the Mediterranean must experience a deficit of water, restored by inflow from the Atlantic through the Straits of Gibraltar. This is probably the earliest example of a mass balance calculation.

Halley also considers the salt cycle on the Earth. Anticipating the 19th century geology view that "the present is the key to the past", he notes[12] that the seas are salty and that the rivers are less so. But while the water transported into the seas is evaporated, to complete the cycle by raining back onto the land and driving the rivers, the salt must accumulate and that there should be a progressive increase in the saltiness of the sea[13] (indeed, one might cheekily suggest such a trend as 'secular' – in this paper Halley grapples with the Age of the Earth inferred from scripture, noting the difficulty of taking the 'seven days' indicated in Holy writ literally on this question). Halley then proposes systematic measurements of the saltiness of lakes and seas. Recognizing that detecting a measurable increase might demand an interval much longer than his own lifetime, he proposes that at least such a survey could act as a baseline for a future comparison, rueing that the Ancient Greeks failed to record such quantitative data for the benefit of such a calculation!

Halley also made innovations in the presentation of data. He compiled information from a variety of sources into a map of the global trade winds - perhaps the first vector map[14]. He similarly mapped the declination of the compass needle, the resultant graphic is considered the first contour plot in the scientific literature. While we take these portrayals for granted today, we should consider that it is only once real measurements are put on a map this way that the workings of our world on a planetary scale can be understood.

As an island nation, England in the Seventeenth (and Britain as a whole in the Eighteenth) century relied for her prosperity and security upon seapower. Geophysics and meteorology (at the time, of course, both disciplines that fell under the umbrella of 'natural philosophy') were therefore of strategic interest in connection with navigation at sea. Innovations in compasses, telescopes, and famously in clocks, were all stimulated by seafaring.

Halley's synthesis of reports from as many sources as he could find into a unified picture of the trade winds is probably his most well-known meteorological contribution. Interestingly, Halley recognizes that his big picture is just that, an idealized approximation of the sea winds at some distance from the land. He recognizes that terrain can cause substantial perturbations to the winds: *'upon and near the shores, the Land and Sea Breezes are almost everywhere sensible; and the great Variety which happens in their Periods, Force and Direction, from the situation of the Mountains, Vallies and Woods, and from the various texture of the Soil, more or less capable of retaining and reflecting Heat, and of exhaling or condensing Vapours is such, that it were an endless task, to endeavour to account for them.'* Of course, the modern approach to climate prediction and weather forecasting is to do exactly this, to divide the world up into tiny boxes and assign estimates for just these effects. In modern terms, Halley is saying a good climate model should include fields of topography, albedo and thermal inertia, well-established for the Earth and now Mars, although we are still catching up on these for Titan and Venus.

A successor of Halley, George Hadley, offered a dynamical explanation of why the trade winds should blow the way they do (1735), recognizing that warm air rises at the equator, where the solar heat is strongest, and so surface air from higher latitudes moves in to replace it. However, the eastwards motion of this air is less than that of the Earth's surface at the equator (which is furthest from the Earth's axis of rotation) and so the wind appears to come from the northwest or southwest rather than purely north or south. Hadley even remarks in his paper that navigation would be 'tedious' without the Earth's rotation, and that winds must balance out over the surface of the globe, otherwise the rotation of the Earth would be affected.

Contemplation of other worlds began to move on, with astronomer Giacomo Maraldi (the nephew of Cassini) observing the poles of Mars in 1704 and suggesting, in 1719, that they are ice caps. A rather curious coincidence of fiction and fact is the speculation by Jonathan Swift in the 1727 adventure 'Gulliver's Travels' that Mars has two moons (which it has – although they'd be impossible to see with the telescopes of the time).

Around this time, the measurement of heat became systematized with the adoption of standard scales of temperature, by Gabriel

Fahrenheit (1724) and Anders Celsius (1742). Once standard temperature scales began to be used, it became possible to compare temperatures recorded in different places. The German astronomer Tobias Mayer[15] developed what might be called the first parametric model of climate around this time, devising mathematical expressions to suggest what the temperature of a given place should be as a function of its latitude, and elevation. He even, in essence a methodology like that used by astronomers to predict tides, suggested a simple sinusoidal function to predict how much later the warmest day should be after the summer solstice.

The French geophysicist Jean-Jacques de Mairan[16] gathered as many temperature measurements as he could − in mines as well as on the surface, and noting that water from springs could be much hotter than the air, deduced that the interior of the Earth must be warm.

Mairan's colleague Buffon[17] attempted to calculate how long the Earth might have taken to cool from an initially hot state by extrapolating from an experiment : he placed iron spheres of different sizes (up to 5 inches) in a fire and observed how long they took to get cold enough to touch, and extrapolated to the size of the Earth, yielding an impressively precise estimate of 96,670 years. Flawed though the calculation was, and indeed the assumption on which it was founded, it was at least a very long time, longer than many had contemplated at the time.

Mairan wondered at how the varying obliquity of light in summer and winter controlled temperature, much as Halley had. Even though there is no direct sunlight at all at the polar regions in winter, they do not become unimaginably cold. Mairan reasoned that perhaps heat from the Earth's interior moderated the temperatures, a paradigm also considered in a more quantitative way by Fourier a century later. Mairan admitted that he could not quantify the effect of the atmosphere, although his colleague Pierre Bouger[18] in 1729 did derive the mathematical law (usually called today the Beer-Lambert law) describing the reduction of light by an absorbing medium.

Some of the first attempts to measure the intensity of sunlight came around this time. Horace-Benedict de Saussure, of Geneva, a geologist and physicist, loved the mountains and enthusiastically and systematically made measurements with barometers (in part to determine the most reliable conditions or time of day to use a barometer to determine altitude) and thermometers, spending some seventeen days at the summit of Col de Geant (3371m) in 1788 recording two-hourly readings. He also devised a reasonably reliable approach to measuring humidity, or the water vapor content of the air : his instrument (a hygrometer) used a human hair as its sensing element. De Saussure experimented with using thermometers in all kinds of different ways, noting the need to ventilate the bulb and shield from sunlight to get a reliable air temperature measurement. He also placed thermometers in glass vessels with a long response time, so that they could be dangled down to the bottom of lakes and then hauled up quickly enough to retain a 'memory' of the lake temperature – he found that lake bottoms consistently had a temperature of 4 degrees Celsius.

De Saussure also invented, in 1767, the first practical solar oven – an insulated box with three layers of glass to trap the heat. He found that the maximum temperature he could attain (230 °F) was more or less the same at the summit of Mt. Crammont as it was on the valley floor, 4852 feet lower down, in sharp contrast to the air temperature which was appreciably warmer at lower altitudes. These experiments provided the first insights into radiative heat transfer in atmospheres, the foundation of modern climate models.

The late 1700s also saw advances in chemistry, with elements such as oxygen and sulfur being recognized by Lavoisier. Also of note in this period are developments due to polymath Benjamin Franklin, who not only explored the electrical nature of storms (famously inventing the lightning conductor to protect buildings), but engaged in a wide range of meteorological investigations. In 1755 he wrote a vivid account[19] of an encounter with a dust devil in Maryland, describing how he rode alongside it and attempted an early experiment in weather modification , seeing if his whip would break up the whirlwind (it didn't!).

As US postmaster, Franklin investigated the great difference in trip times of mail ships which took weeks longer to cross the Atlantic coming from England than going to it. Assembling accounts from whalers and ship captains, he assembled in 1770 a chart of the current responsible, which he named the Gulf Stream. This current plays a large role on moderating the climate of northern Europe, transporting heat from lower latitudes. Franklin also suggested a possible link between the 1783 eruption of the Laki volcano, and the harsh winter in Europe the following year.

In fact Franklin was not the first prominent American to discuss changes in the weather and climate, and indeed the possible influence of human activity on climate. Thomas Jefferson, who as a farmer took a close interest in the land and weather, kept a diligent record of weather, and reduced his records into averages. He noted that conditions were warming : "*A change in our climate however is taking place very sensibly. Both heats and colds are become much more moderate within the memory even of the middle-aged. Snows are less frequent and less deep. They do not often lie, below the mountains, more than one, two, or three days, and very rarely a week. They are remembered to have been formerly frequent, deep, and of long continuance. The elderly inform me the earth used to be covered with snow about three months in every year. The rivers, which then seldom failed to freeze over in the course of the winter, scarcely ever do so now. This change has produced an unfortunate fluctuation between heat and cold, in the spring of the year, which is very fatal to fruits".* [20]Jefferson speculated that clearing of the East coast forests by settlement now allowed the sea breezes to penetrate further inland, moderating the climate[21]. Jefferson also makes an observation recognizing the difference between air temperatures and ground temperatures, *"The access of frost in autumn, and its recess in the spring, do not seem to depend merely on the degree of cold; much less on the air's being at the freezing point. White frosts are frequent when the thermometer is at 47 degrees, have killed young plants of Indian corn at 48 degrees, and have been known at 54 degrees."* Jefferson even speculated, in what must be one of the most early instances of contemplated planetary engineering, what the change on climate might be if the Spanish were to cut a channel in the isthmus of Panama wide enough to allow tropical ocean currents to pass, perhaps eliminating the fog banks off Newfoundland.

A remarkable compilation of climate information ('An Estimate of the Temperature of Different Latitudes') was published in 1787 by Richard Kirwan, an Irish geologist and chemist. He reviews the ideas of Mayer and de Mairan (having spent 19 years in London, he maintained a vigorous correspondence with many scientific luminaries of the time) and tabulates Mayer's calculations. More usefully, he also compiles the measurements he obtained from his contacts, from Kamchatka and Algiers, to Labrador and Quito, the Cape of Good Hope and Tibet. Although the exercise is descriptive rather than predictive, Kirwan assesses various factors influencing the weather (e.g. whether a cool summer preceded a cold winter, and the role of Siberian winds) notes that journals implied that winds measured in America were subsequently encountered in Europe, an early correlation study. He also confronts the puzzle of interannual variability yet offers hope that the scientific method may eventually crack the problem – "*Since the astronomical source of heat is permanent, and the local causes of its modification undergo no annual variation, and yet the temperature of no two succeeding years is perfectly alike, it is evident that this annual variation proceeds from causes equally variable. Of these, there may be many, but at present we know of none, that have a demonstrable influence on the weather, but winds' and since winds themselves, however uncertain in appearance, are like all the other phenomena of nature, governed by fixed and determinate laws, they deserve the most serious investigation, for which we are at present, tolerably well-prepared.* "

2

PLANETS AND GREENHOUSES

1780 to ~1870

Developments in astronomy moved somewhat slowly in this period. But special geometrical events allowed some progress : the first indication that Venus had an atmosphere came in 1761, when the Russian astronomer Mikhail Lomonosov observed the transit of Venus across the disc of the sun. He noticed (using a small refractor telescope with a smoked glass filter to suppress the light) that the 'tidy' edge of the sun was distorted by Venus' passage, and particularly that a bulge in the solar disc was caused by refraction of the sun's light by an atmosphere. Lomonosov also argued that the Earth must have a southern polar continent (i.e. Antarctica) because icebergs seen in the southern oceans must have formed on land.

Similar observations a quarter century later showed to the German/British astronomer William Herschel that the Martian atmosphere was different from ours, in his 1784 paper entitled *"On the remarkable appearances at the polar regions on the planet Mars, the inclination of its axis, the position of its poles, and its spheroidal figure"*. When two faint stars passed close to Mars without showing variation in brightness, Herschel inferred that the Martian atmosphere had to be thin. Herschel also noticed that the south polar cap on Mars wasn't quite centered on the geometric pole, which he saw to be inclined to Mars' orbit around the sun much the same as the Earth's, meaning Mars had seasons much like Earth.

One of Herschel's most important discoveries, in 1800, was not made through a telescope, however. He passed sunlight through a glass prism to make a rainbow, and placed blackened thermometers in different colors to measure their temperature, in effect measuring the power in each wavelength of light. He found that the temperatures increased from the violet to the red end of the spectrum, but noticed that a position just beyond red (where there was no visible light at all) had the highest temperature. He inferred there must be invisible radiation which he called "calorific rays" which were reflected, refracted and absorbed just like light, but with a longer wavelength : what we now call the infrared.

Herschel also observed the sun, and sunspots in particular, which had been noted by Meton of Athens in the 5th century BC, as well as by Galileo and many others after. Strikingly, in 1801 he noted[22] that the price of wheat varied (as documented in Scottish economist Adam Smith's "Wealth of Nations") over the previous century in a manner that seemed to relate to the number of sunspots (which he assumed might be associated with a drop in solar output).   His discussion highlights the fundamental challenge in climate change studies, that of attribution : correlation does not guarantee causation : *"The result of this review of the foregoing five periods is, that, from the price of wheat, it seems probable that some temporary scarcity or defect of vegetation has generally taken place, when the sun has been without those appearances which we surmise to be symptoms of copious emission of light and heat. In order, however, to make this an argument in favour of our hypothesis, even if the reality of a defective vegetation of grain were sufficiently established by its enhanced price, it would still be necessary to shew that a deficiency of the solar beams had been the occasion of it."* This work embraces three important ideas[23] – that the climate changes, that the sun changes (contrary to Kirwan's assumption), and that the sun might affect climate. Herschel ventures to suggest on the basis of the sunspot cycle that harvests in the coming years should be good, but cautions that *"the subject, however, being so new, it will be proper to conclude, by adding ,that this prediction ought not to be relied on by any one."*

About this time, a Scottish farmer and scholar, James Hutton, had developed a perspective of the Earth as undergoing constant change. His *"Theory of the Earth; or an Investigation of the Laws observable in the Composition, Dissolution, and Restoration of Land upon the Globe"* was read to the Royal Society of Edinburgh in 1785. He recognized that much of the geological record was shaped by processes that can be observed in action today, and thus that these processes must have acted for a considerable period to form the landscape we see today, with layer upon layer of sediments laid down underwater, heated and turned to rock and sometimes thrust back up as mountains, only to be worn down again. This ongoing cycle saw *'no vestige of a beginning, no prospect of an end'*.

Hutton, a close friend of fellow Scot Joseph Black (who discovered latent heat, and identified carbon dioxide as 'fixed air'), also developed a Theory of Rain, reasoning that since the amount of water vapor that could be held in the air increased non-linearly with temperature, that mixing of warm and moist air with cooler air would prompt condensation and thus rain. The properties of water vapor and heat were of intense interest in the development of the steam engine at this time; Black helped influence, and finance, James Watt's experiments.

Chemistry emerged as a discipline, and scientists began to understand that the Greek element 'air' was a mix of different compounds, and with careful weighing, the French chemist

Antoine Lavoisier deduced that 'fixed air' was about 25% by weight carbon and 75% of what had been called 'vital air' or 'dephlogisticated air', but what he named 'oxygen'[24]. Plants and animals exchanged these gases : animal respiration in particular was analogous to combustion, drawing in 'vital air' and exhaling 'fixed air', which dissolved in water to make a weak acid, which Lavoisier called 'carbonic acid'. We now call the gas (knowing the relative amounts and atomic weights) carbon dioxide – two atoms of oxygen to one of carbon. Carbonic acid had also been called 'chalky acid' – indeed in the laboratory, pouring acid on limestone, chalk or other carbonate rocks was the easiest way to make pure carbon dioxide, just as adding acid to metals like iron was a way to make the gas 'hydrogen'.

Another notable French development in this period was that of the balloon – using hot air by the Mongolfier brothers in 1783, and using hydrogen gas that same year by Jacques Charles[25]. These developments paved the way for the exploration of the atmosphere, and Joseph Gay-Lussac in 1804 ascended to 7 km, recording temperatures from which he deduced a steady fall of temperature with altitude. Later with Alexander von Humboldt (then resident in Paris) he determined that the composition of air did not change measurably with height, and that water was made of two parts hydrogen and one of oxygen.

Humboldt, a vigorous Prussian naturalist, had explored South America extensively. Among his discoveries were the electric eel in the Amazon, and the cold ocean current off Peru that bears his name. He was an enthusiastic mountain-climber, reaching 5,800m on the Andean volcanic peak Chimborazo in Ecuador, a world record at the time. Humboldt[26] wrote as widely as he travelled, and he documented how elevation produced a change in local climate that mirrored changes in latitude (e.g. the climate at sea level at 30° latitude was the same as that at the equator at 1500m elevation), which he mapped out to develop a picture of the terrestrial climate overall. The power of graphics in scientific communication is once again underscored – although Kirwan, de Mairan and Mayer had all quantified variations with latitude and altitude, it is Humboldt that usually gets the credit, for laying these variations out in a readily-assimilated way.

The invention and standardization of the barometer and thermometer enabled quantitative observations, long term climate records. Further, in 1803 an English chemist and amateur meteorologist, Luke Howard, proposed a systematic classification of clouds into three principal categories of clouds – cumulus, stratus, and cirrus – and their combinations[27]. Howard also compiled measurements into *The Climate of London* 1818-1820, a first review of urban climate. He noted 'city fog' (or what we would call smog) and that the temperatures in the city at night were several degrees warmer than simultaneous measurements made in the nearby countryside, an important indication of the human impact on climate. Another English chemist at this time, John Daniell, best-known for his design of an electric cell, developed an measure of humidity, the dew-point hygrometer : by expressing moisture as a temperature, this method was easy to standardize in contrast to the catgut and hair instruments used before. In 1824 Daniell also wrote an *Essay on Artificial Climate considered in its Applications to Horticulture*, documenting the importance of enhanced humidity in greenhouses.

Back in France, Joseph Fourier in 1822 calculated that the Earth would be far colder if it lacked an atmosphere, and notes (in papers in 1824 and 1827) that underground temperatures do not show diurnal variations, but do show a variation with latitude. He recognizes that this varation with latitude is less than pure geometry of sunlight would produce, but must be moderated by the transport of heat in the atmosphere and oceans. He even recognizes anthropogenic climate change "*The movements of the air and the waters, the extent of the seas, the elevation and the form of the surface, the effects of human industry and all the accidental changes to the terrestrial surface modify the temperatures in each climate.*"

Much has been written[28], in connection with the history of understanding the politically-charged greenhouse effect, about what Fourier does and does not say in his paper, so let's spend a little time on it here. My impression – emphasizing the planetary perspective - is that he gets a lot of things right, and he acknowledges what is not known for sure.

First, he outlines that the temperature of the earth depends on three reservoirs of heat – the sun, the interior of the earth, and the temperature of planetary space. We'll come back to the last one. Fourier (whose eponymous decomposition of functions into sums of sinusoids is useful in problems of thermal conduction) recognizes that the heat leaking out from the interior of the Earth has a miniscule influence on our present climate.

He recognizes that he cannot quantify the effect of the atmosphere on terrestrial temperatures, but draws attention to measurements by Swiss physicist and alpine geologist Horace-Benedict de Saussure of temperatures in an insulated box with a glass window which he brought to different mountain elevations. The contemplation of these experiments and similar analogies led to the term 'greenhouse effect' – we leave aside here how formally accurate an analogy to an atmosphere it is, since the suppression of convection and the absorbance of thermal radiation are different things, but it is the warming-by-blocking effect that matters here. He also recognizes the distinction between 'luminous' and 'non-luminous' heat – in essence what modern climatologists would call visible and thermal radiation, or short-wave and long-wave radiation. Further, he understands that materials may allow these two streams of energy to pass in different degrees. So, he certainly sets the stage for understanding the greenhouse effect – as he puts it *'luminous heat flowing in penetrates...and non-luminous heat has more difficulty in finding its way out.'*

He doesn't quantify how much heat is moved around the earth by the atmosphere and oceans, but notes that these motions do attenuate the difference in temperature between equator and pole. Not having an estimate, he then (it is a slightly rambling and self-contradictory paper, worth reading nonetheless) more or less argues that the poles of the earth should get unimaginably cold, since they get very little heat from the sun. So they must get some heat from space – arguing that either the temperature of space around the Earth, or the effect of starlight – keeps the poles from getting too cold. In effect he imagines space as a heat sink *'the influence of stars is equivalent to the presence of an immense hollow sphere, with the Earth in the center, the constant temperature of which should be a little below what would be observed in the polar regions'.* He's wrong here[29], but the picture fits the data. He suggests that the other planets cannot be much colder than our poles, since even though they get less sunlight than we do, they are still immersed in this same heat sink of space.

Impressively, while explaining how he gets a picture mostly consistent with what is observed, Fourier admits (of the other planets) *'Now, they would be entirely different if we were to admit an absolute cold in space'* (which is in fact the case). So Fourier gets the overall picture just about right, although there isn't enough known about atmospheric opacity and air motions transporting heat to distinguish their effects from those of (also unknown) space.

Interestingly, around the same time as Fourier's memoir, his countryman Sadi Carnot, making efforts to understand the fundamentals behind the performance of steam engines, opens his treatise 'Reflections on the Motive Power of Heat', with the solar heating/air motion causality in the other direction : '*To the agency of heat may be ascribed those vast disturbances which we see occurring everywhere on the earth ; the movements of the atmosphere, the rising of mists, the fall of rain and other meteors[30], the streams of water which channel the surface of the earth, of which man has succeeded in utilizing only a small part*'. Carnot saw atmospheres, correctly, as engines. We will return to this useful perspective later. The following decades saw the formulation of the more formal laws of thermodynamics : the concept of entropy by Rudolf Clausius, and he and Benoit Clapeyron laid out a basis for the vapor pressure of materials such as water, of critical importance to climate. William Thomson, later Lord Kelvin (the first scientist to become member of the House of Lords in the British Parliament) took their work further and devised the absolute scale of temperature[31] that bears his name in 1848.

Kelvin in 1864, armed with a better understanding of heat conduction, and with measurements of the thermal conductivity of rocks and temperature gradient in the Earth's crust, deduced a cooling age for the Earth of tens of millions of years. Among the underground temperatures bearing on this were a set measured with special thermometers installed in 1837 at depths of 3, 6, 12 and 24 feet in the porphyry rock of Carlton Hill in Edinburgh, Scotland. These instruments had long tubes with scales at the surface, and were read once a week for the next sixteen years, with a claimed precision of a hundredth of a degree Fahrenheit. Scottish astronomer Charles Piazzi-Smyth[32] examined this formidable dataset in 1856, noting how the thermal wave of summer propagated into the rock, with the 3-ft wave with a 15 degree amplitude peaking in August, but at 24-ft a mere 1.2 degrees, peaking in January[33]. The mean temperatures also showed a consistent increase with depth, of about 1 Fahrenheit for 21 feet of depth. Piazzi-Smyth then attempted to subtract all these effects out, to see what variation was left, in the hope of detecting some interannual variation which might be due to the sun. He plotted the temperature variation (which was pushing the quality of the data rather hard) against observations of sunspots, but concedes 'no very near approach can be claimed to the temperature curves'.

That same year, Piazzi-Smyth set off on an expedition to explore the question of how much better an astronomical observatory might be by being placed on a mountaintop. He sailed to Tenerife, a volcanic island off the west coast of Africa, armed with a portable telescope (portable, that is, by four mules!) and a variety of meteorological instruments. His team set up on the flanks of Guajara, at an elevation of about 9000 ft. Although this height was well above the usual cloud-deck and skies were often beautifully clear, Piazzi-Smyth noted (in his book[34], which is remarkably illustrated with "photo-stereographs") and interest that astronomical conditions were sometimes degraded by a thick haze of dust (which had blown in from the Sahara desert, some 300 miles to the east). When conditions were good, however, they were very good, and the views of Jupiter allowed him to note the calm nature of its poles with cirrostratus clouds, while the tropics had tempestuous cumulostratus : he observed that the "medial line of calm" was not exactly coincident with the equator, and wondered if the causes were the same (greater proportion of land area in the northern hemisphere) as why the calm between the northern and southern trade winds was pushed north of the equator on Earth, towards the latitude of Tenerife.

By day, Piazzi-Smyth noted the tough conditions at altitude: the varnish on his barometer (which read 22 inches at the Guajara station) had blistered in the harsh sunlight after only one day. He observed how dry the air was, the dew point having dropped by some 40 degrees in their ascent – *"No wonder we felt our lips cracking, our hair frizzling, our nails becoming brittle, and saw each other's faces scarlet."*

Piazzi-Smyth's team made measurements with solar thermometers in the thin air, making the observation that would become a familiar feature of the Martian climate, namely dusty days (when the solar thermometer's reading was reduced by attenuation of sunlight by the dust) saw the air temperature in the shade being elevated. Piazzi-Smyth even describes being blasted by pebbles and blinded by dust in a dust-devil, a whirlwind that came out of nowhere – as they often do on Mars. At the time the generation of hurricanes and other rotating storm systems was not understood, and it was even speculated that electricity might somehow be responsible – Piazzi-Smyth pondered hurricanes after his dust devil encounter, noting that even the dust devil *"required an exertion of a considerable amount of mechanical energy; and the question to be settled, is, which of the two elements present on the occasion, was most capable of producing the effect observed, wind or electricity?"*

P. Baddeley, a surgeon in the (British) Bengal Army, confronted the same question[35] around 1850, whether electrical phenomena were the cause, or the result, of whirlwinds. In India, he chased dust devils, observing the behaviour of a gold-leaf electroscope, and noting that a paper moist with starch and iodide solution became discolored, indicating the presence of ozone. (Ozone itself had only been isolated in 1839, and was determined to be a triatomic molecule of oxygen in 1867).

The nature of storms was a matter of considerable interest in the 1800s, as it has always been, but getting a wider picture of weather systems was somewhat easier on the expanding continental United States than in fractured Europe, where storms tended to arrive from the vast and unobserved Atlantic. William Redfield, a New England saddle-maker, noticed that the alignment of fallen trees after a major storm in 1821 differed from place to place, and reconstructed the storm's path from newspaper reports and interviews with sailors. He advanced the idea that winds swept circularly around storms – they were large whirlwinds. A contrary view was vigorously espoused by James Espy, who argued that winds converged towards storms, and rose in the center. Of course, each proponent had truth partly on their side, so while the debate was not immediately resolved, it stimulated the acquisition of better data. The rapidly-expanding telegraph network allowed weather systems to be reported before they arrived, and Joseph Henry, an electrical engineer and director of the Smithsonian Institution in Washington, D.C. arranged for telegraph operators to provide brief routine weather reports[36]. By 1858, Henry had a large map on display in the lobby of the Smithsonian, displaying weather reports telegraphed that day from as far as Iowa and Missouri, as near to a real-time global display as the 19th century got. Another party in the debate was mathematician William Ferrel, who understood more clearly than others how the Earth's rotation influenced air masses[37], and elaborated on Hadley's ideas. Ferrel also recognized the structure of the vertical and meridional (north-south) circulation of the atmosphere, with upwelling at the equator and downwelling near 30 degrees latitude (which is termed the Hadley cell), a midlatitude cell rotating in the opposite sense, named after Ferrel, and a polar cell. The position of the downwelling depends on planetary rotation, and is quite different on other planets.

As the physical nature of heat was becoming understood in the laboratory (and, indeed, the brewery[38]), outdoors the realization was emerging that the Earth's climate was not invariant. In particular suspicions grew, most vigorously propounded by Swiss Louis Agassiz circa 1837, that much of Europe had once been covered by a great sheet of ice, perhaps a mile thick. This staggering idea could explain many features of the recent geological record, such as scratches on rocks, and the presence of large erratic boulders hundreds of miles from their source (in fact Hutton had speculated ice might have been their cause). But how could the climate have changed so dramatically ? In 1842 Joseph Adhemar recognized that the Earth's elliptical orbit leads to slight asymmetry in the seasons, and he suggested in his book 'Revolutions of the Sea' that astronomical change (specifically, the 26,000-yr precession of the equinoxes, known even to ancient astronomers) might be responsible.

Meanwhile, the world was becoming smaller. Not only were voyages of exploration like Humboldt's, and that of H.M.S. Beagle in 1836, carrying Charles Darwin, bringing a more complete understanding of the world and its inhabitants, but communications were fast improving. First railway networks, and then the telegraph, meant that information could travel faster than a horse. It now became possible to compare weather conditions at different places simultaneously, to literally develop a picture of the atmosphere.

While observations at different places of a few storms began to be assembled into weather maps[39] (after the fact) in the 1820s and 1830s, it was James Glaisher – appointed in 1839 as the Superintendent of the Magnetic and Meteorological Department at the Royal Observatory in Greenwich, England who first systematically began assembling such observations (brought by railway) and his consolidated reports of the weather yesterday began to be published in newspapers in 1849.

Matthew Maury, an ambitious oceanographer with the US Navy, saw that to take the next step, to anticipate or forecast the weather, would demand international collaboration, and convened an International Meteorological Conference in Brussels in 1853. In the following year, a dedicated British government department was formed, essentially the forerunner of the modern Met Office, led by the Beagle's former sea captain, Robert Fitzroy (France's meteorological service was set up by the astronomer Urban Leverrier at the Paris Observatory the same year). Fitzroy set up stations with standardized instruments, and had them send the readings several times a day via telegraph. This allowed him to issue storm warnings to ports when violent conditions were expected. It was only a small step further to report conditions when no storms were expected, and the first regular weather forecasts began to be published in 1861. Because of its strategic importance, in many countries, the weather reporting function was seen as part of the military – for example early US weather maps were published by the Signal Service of the War Department.

Although the day-to-day variations in weather were now being tracked and progressively better-understood, scientists were still grappling with the factors that controlled the long-term conditions. In particular, the effects of the atmosphere on sunlight and heat had to be quantified. A paradox was the variation of temperature with altitude – how could it get colder when you got closer to the Sun? Glaisher made a famous ascent in a gas balloon in 1862, carrying instruments to measure pressure, temperature and humidity, and found that Gay-Lussac's altitude trend did not persist indefinitely – these balloon flights gave the first inklings of the existence of the stratosphere. In fact, Glaisher lost consciousness during his flight, which inadvertently reached above some 30,000ft, the lower reaches of the stratosphere – he was only saved by his pilot Henry Coxwell who managed to pull a gas vent valve with his teeth, his hands rendered useless by frostbite.

The role of humidity in the atmosphere turned out to be crucial to the climate question. While the blanketing effect of the atmosphere was long-suspected, it was John Tyndall[40] who demonstrated in 1959, using a thermopile heat detector, the blocking of radiant heat by certain gases, notably water vapor and carbon dioxide. Tyndall, a great communicator as well as experimentalist, suggested that changes in the concentrations of these gases could alter climate with the following analogy "*As a dam built across a river causes a local deepening of the stream, so our atmosphere, thrown as a barrier across the terrestrial rays, produces a local heightening of the temperature at the Earth's surface.*"

Tyndall's progress arose in part from the exquisite sensitivity of his apparatus : he set it up not to measure the absolute amount of caloric rays transmitted by the gas (held in a metal tube with rock-salt windows – glass is opaque to infrared radiation, so the 'greenhouse effect' is not a complete misnomer), but to measure the difference between that and a reference, which he could adjust to be equal to that transmitted by the gas tube evacuated of gas. Measuring this difference, with a delicate galvanometer[41], let him detect much smaller absorptions. The Italian Macedonio Melloni used a similar detector, at the focus of a 1-m wide Fresnel lens, to make the first detection[42] of the heat of the moon in 1845.

Tyndall's experiments showed that the ability of gases to emit infrared radiation was directly related to their ability to absorb it. He also showed that the amount of absorption increased proportionally with the amount of gas – at first – but at larger amounts, the absorption levelled off.

He determined[43] that the water vapor in the atmosphere was the major factor – in his experiment causing 13 times more absorption than the other gases. He concluded "*every variation of this constituent must produce a change of climate. Similar remarks would apply to the carbonic acid diffused through the air; while an almost inappreciable admixture of any of the hydrocarbon vapours would produce great effects on the terrestrial rays and produce corresponding changes of climate."* A little over a century later, planetary scientists would invoke such hydrocarbon vapors to solve climate puzzles, but of course carbonic acid (carbon dioxide) would occupy center stage on Earth.

It was only at this time realized that invisible radiations that caused chemical reactions ('actinic' or what we'd now call ultraviolet), visible light, and thermal radiation (infrared) were the same thing – Melloni, for example realized only in 1842, *"The dark undulations responsible for chemical or calorific actions are perfectly similar to luminous undulations; they only differ from them by wavelength."* Although it is the blockage of long-wavelength ("thermal", longward of say 3 microns) infrared radiation from the Earth's warm surface that is responsible for our greenhouse effect, the distinction between this flavour of invisible infrared light, and another (what we'd call near-infrared, between about 0.8 and 3 microns, and abundant in sunlight) took some time to be appreciated.

The absorption of gases of the latter radiation was documented in a short paper[44] a couple of years earlier than Tyndall, by Eunice Foote, a notably early American woman scientist. By simply setting out two glass vessels with thermometers in the sunlight, she determined that the warming varied with the density of the air (using a pump to increase the air density in one and decrease it in the other). Furthermore, she measured that the warming was greater in moist air than dry air, and found that the effect was particularly strong with 'carbonic acid gas' – strong enough to detect by hand without even using the thermometer. She speculated *'An atmosphere of that gas would give to our Earth a high temperature'.*

The possibility of variations in the amounts of carbon dioxide and oxygen in the atmosphere was presciently outlined by French chemist and mining engineer Jacques-Joseph Ebelmen who wrote in the 1840s[45] *"I see in volcanic phenomena the principal cause that restores carbon dioxide to the atmosphere that is removed by the decomposition of rocks"* – his countryman Boussingault in the 1830s had sampled the gas from several volcanoes and found carbon dioxide to be a major component,[46] whereas the Swiss Nicolas de Saussure (the son of Horace-Benedict, whose glass-box experiments had inspired Fourier's thoughts on the greenhouse) showed in 1804 that atmospheric air contains only a tiny amount. Meanwhile, Ebelmen found carbon dioxide was removed by reaction with silicate rocks, a process (sometimes referred to in modern times as the 'Urey' reaction) usually mediated by the action of plants, and then its ultimate sink was in carbonate mud in the oceans - *"The terrestrially-derived carbonates end up by being deposited or they are taken up by marine animals"*. Without the benefit of Tyndall's measurements, he qualitatively guessed correctly *"... in ancient geologic epochs the atmosphere was denser and richer in CO_2, and perhaps O_2, than at present. To a greater weight of the gaseous envelope should correspond a stronger condensation of solar heat"*. Ebelmen introduced the phrase "carbon rotation", noting that carbon dioxide in the air might be taken up by plants (and, as Hutton and many others thereafter recognized) those plants might be turned into coal, the carbon locked up in the ground, or made into the carbonate shells of marine life and perhaps into limestone, and then the carbon might be released again by combustion or volcanic heating of carbonates. The modern carbon cycle emerged.

Meanwhile, an impressively multidisciplinary approach to the question of the ice ages was doggedly made in the spare time of a Scottish janitor, James Croll, who after several unsuccessful careers in other fields enjoyed his job at the Andersonian College and Museum in Glasgow because it allowed him use of its library. Croll published a short paper in August of 1864 estimating the changes in the Earth's climate that might be due to changes in orbital eccentricity. His subsequent work on the ice-age problem was to occupy his attention for the next twenty years.

Croll recognized that the changes in sunlight due to these astronomical changes were small, but suggested one means by which they could be amplified. He noted that snow-covered ground (or ice-covered seas) would reflect more sunlight than their bare equivalents, and so slightly colder conditions could allow a bit more terrain to be snow-covered for longer, and would absorb a bit less heat as a result, allowing snow to persist a bit longer again. Croll introduced the idea of a climate feedback (specifically, the 'ice-albedo' feedback.)

Croll was perhaps the first person to calculate how much colder regions at high latitude would be if there were no heat transport from lower latitudes, recognizing that high latitudes receive so little sunlight (as originally calculated by Halley) that the contribution of ocean currents and winds dominates. He calculated that London was 10 degrees warmer than the average temperature at that latitude, and that the latitude was on average 30 degrees warmer than it should be from sunlight considerations alone. He noted that the Gulf Stream provides some 77,479,650,000,000,000,000 foot-pounds of energy to the Atlantic, or roughly one-fourth of the heat received by the sun.

The oceans were a hot scientific topic at the time, with amazing deep-sea creatures among the 4000 species discovered by H.M.S. Challenger. Challenger's epic expedition 1873-1876 made hundreds of soundings of temperatures and salinity at depth, as well as sampling seabed sediments. The tens of thousands of pages of the expedition's findings took some 19 years to be fully documented, and its deep-sea temperature measurements are still referenced today[47].

Croll's researches were summarized in an 1890 book with the grand title 'Climate and Time' (more fully ; 'Climate and Time in their Geological relations: A Theory of the Secular Changes of the Earth's Climate'). Although he got the big picture right, his calculations on temperature were based on the assumption that the temperature was proportional to the supplied heat (both from the sun, as a function of the changing orbital characteristics and latitude, and from ocean currents), whereas the correct dependence was only at this time being worked out by physicists, primarily in Germany and Austria.

However, Croll not only considered the astronomical aspects of the problem, calculating at length the influence of changes in the Earth's orbital eccentricity and obliquity (i.e. the tilt of its spin axis) on the amount of sunlight at different latitudes and seasons, but he also explored the geological evidence for glaciation. In particular he investigated the widespread boulder clay, and the shells of arctic creatures now extinct in interglacial Scotland. In the century to follow, his multidisciplinary approach would become rare as science became more and more specialized. The end of this epoch, in the last decades of the 1800s, marks the onset of ever-more quantitative methods in science, and, at last, the accurate evaluation of conditions on other worlds.

3

CALCULATION OF CLIMATE

1870 TO 1930

The late 1800s saw the horizons of popular imagination venturing out to the planets. Jules Verne, of course, wrote in 1865 the novel From the Earth to the Moon, wherein the Baltimore Gun Club established (in Florida, presciently enough, not far from the modern Kennedy Space Center) a giant cannon to launch a human crew to the moon. H. G. Wells' 1898 science fiction 'War of the Worlds' imagined interplanetary transportation in the other direction, with Martians invading southern England in cylinders whose gun-launch from Mars was visible in the telescope. Interplanetary gunnery was not especially practical, as even Huygens had noted. But In 1867, a Scottish clergyman, William Leitch, wrote a review[48] of what was known about the solar system, and noted that *'the only machine, independent of the atmosphere, we can conceive of, would be one on the principle of the rocket. The rocket rises in the air, not from the resistance offered by the atmosphere to its fiery stream, but from the internal reaction. The velocity would, indeed, be greater in a vacuum[49] than in the atmosphere....we might, with such a machine, transcend the boundaries of our globe, and visit other orbs'.*

While the temperature and atmosphere on Mars was not yet known, circumstantial evidence against its habitability began to emerge. Although Giovanni Schiaparelli , observing Mars in its 'Great Opposition' in 1877, had sketched a map with dark lineaments that he called 'canali' (channels – mistranslated as 'canals'), he was himself doubtful that they were seas, noting that smooth liquid surfaces would reflect the sun like a mirror. The fact that these 'specular reflections' were not observed in telescopes meant that there were no smooth seas. This point was elaborated by Dennis Taylor[50] in 1894, who suggested that surface waters would at least sometimes be smooth, and the absence of recorded sunglint meant that widespread liquid could not be present. He also suggested that the rapid disappearance of the seasonal caps meant they could only be a few yards thick.

Progress in thermodynamics, and in particular in understanding radiant heat, finally laid the quantitative basis for understanding planetary temperatures. Jozef Stefan in Vienna deduced in 1879 from Tyndall's measurements that the heat flux radiated by an object varied as the fourth power of its absolute temperature, and Ludwig Boltzmann derived the result theoretically 5 years later. Stefan even calculated the temperature of the sun, using an observation that a metal plate at about 2000 K radiated about 29 times less heat than did the sun. Guessing that about a third of the sun's flux was absorbed by the atmosphere, meant the temperature ratio had to be about $(1.5 \times 29)^{0.25}$ or ~2.6 times that of the plate, or about 5700K, quite close to the modern value.

It takes some detective work to find exactly who first used the formula to estimate planetary temperatures: in a 1903 paper[51], John Poynting attributes it to Wilhelm Wien circa 1901, but in a footnote in a later paper credits the Danish scientist Christian Christiansen[52] in 1885 with computing a table of planetary temperatures. Specifically, Christiansen (writing in a Danish journal) notes that the radiated heat varies as the fourth power of absolute temperature, and the radiated heat must balance that received from the sun, which varies as the inverse square of distance. Thus the absolute temperature varies as the square root of distance. Christiansen's paper lists the temperatures computed using the mean distances of the planets : Venus would be about 60°C, while Mars was -40°C and Saturn and therefore Titan would be -180°C.

The deep cold of the outer planets, recognized qualitatively by Huygens, is evident in these computations, as is the unpromising chill of Mars – about which much more shortly. Venus entertains, perhaps, some hope of being a torrid kind of habitable – but of course its exceptional greenhouse warming of that planet was not known at the time. Christiansen suggests that Mars' conditions were not unlike Greenland.

Around this time, astronomy began to progress, not only from a proliferation of ever-larger telescopes at ever-better sites, but by the substitution of instruments and photographic film, rather than the human eye, at their foci. After a number of plausible but unreproduced attempts, Lord Rosse using the 3-foot reflecting telescope[53] at his estate in Ireland was able to reliably detect the heat from the moon in 1868-69 with a thermopile. The long-wave and short-wave components of radiation from the moon could be separated by interposing a sheet of glass, and Rosse reported that the proportion of long-wave radiation was much higher in moonlight than sunlight. Furthermore, the total brightness varied roughly in proportion with the phase of the moon – in other words, the moon absorbed much of the sunlight falling on it, re-radiating it as heat, and that the moon warmed up and cooled down relatively quickly. Because the relative absorptions of short- and long-wave radiation by the atmosphere was still unknown, a precise temperature could not be derived but Rosse reported that the day/night swing of lunar temperatures was 500 F.[54]

These observations were brought to a higher degree of fidelity by Samuel Pierpont Langley, of the Allegheny Observatory in Pittsburgh, and later Director of the Smithsonian Institution in Washington. He devised a bolometer detector, more sensitive than the thermopile, and using a rock salt prism to spread infrared light, was able to measure the thermal spectrum of the sun and moon (in which we can recognize the absorptions of major gases – ozone, as well as carbon dioxide and water vapor). He was able in the 1880s to measure the temperatures of different parts of the moon, and observe during a lunar eclipse – confirming the *"extraordinary rapidity with which the lunar surface parts with its heat."*[55]

Langley not only measured these absorptions in the laboratory, but sought to understand how absorption of sunlight and radiation of heat might vary throughout the atmosphere, and mounted an expedition to Mt. Whitney in California[56], dragging his crew and 2 tons of instruments across the country by train, then (with army escort, the West was still wild..) up to 12,000 ft by mules over summer 1881. Langley's attentions later moved to aviation, his steam-powered model planes launched by catapult from a boat on the Potomac showing early promise, but he was beaten out by the Wright brothers in the race for human flight. Langley's program of research at the Smithsonian into the solar constant and atmospheric absorptions was later taken up by Abbott.

In part drawing on Langley's data, the Swedish physicist Svante Arrhenius in 1896 published[57] the first calculation of changing global climate taking into account the warming by carbon dioxide. This exercise demanded tens of thousands of tedious calculations. With glaciation in mind, he noted that if CO_2 levels halved, then the Earth's surface temperature would fall by 4-5 $^{\circ}C$, taking into account the feedback from water vapor. Arrhenius noted the flipside too : doubling CO_2 levels would trigger a rise of about 5-6 °C. However, he didn't expect this would happen for thousands of years, and (from his wintry perspective in Sweden) if it did, it might be a pleasant change.

At this point, several factors were recognized as potentially contributing to climate change, and the ice ages in particular. First, the astronomical variations suggested by Adhemar and Croll : a challenge here being that glaciation in the northern hemisphere should alternate with glaciation in the south. Volcanic perturbations to the atmosphere caught wide attention after the 1883 eruption of Krakatoa, which was perhaps the first globally-recognized geophysical event, the explosion being picked up by barographs (recording barometers) worldwide. The reflective sulfate aerosols injected into the stratosphere caused notable cooling in the following couple of years, and the dramatic light scattered near sunset may have motivated the bloody skies in Edvard Munch's painting 'The Scream'. And then the possibility that carbon dioxide levels might have varied due to vegetation or similar changes. Finally, there was the sun itself which was now known not to be constant in appearance – the 11-year cycle in the number of sunspots was discovered by Schwabe in 1845. Longer-term variations were noticed by another German astronomer, Gustav Spoerer in 1889. His findings were shortly thereafter elaborated by E. W. Maunder in England who noted that there had been a prolonged period 1645-1715 when sunspots were comparatively rare, a period often referred to as the 'Maunder Minimum' and coincident with a period of lower temperatures in Europe, sometimes called 'The Little Ice Age'.

As with many scientific debates, framed in the sense of one cause versus another, all of these factors may be significant. But these purely terrestrial considerations were merely paint on a deeper canvas, that of our sun. Nobody knew how the sun worked. Realizing that the moon shone only because it reflected the sun's light, didn't help explain why it didn't glow in the first place, nor did the findings of Rosse and Langley that it radiated more heat than it reflected light. It was known that the deep interior of the Earth was warm (as Fourier noted) but that the heat leaking out was tiny compared with the sun's illumination – but was that true everywhere? The estimates by Christiansen of Jupiter's and Saturn's temperatures acknowledged that perhaps they could be much warmer if they glowed from internal heat as well as reflected sunlight. The Earth's heat, and its apparent age (many millions of years, to judge from the evidence of Hutton and his geologist successors) could be explained by gravity, the potential energy liberated as smaller bodies accumulated and impacted together to form the planet as a whole. Knowing the heat flow in the Earth, from temperature measurements in deep mines and from the conductivity of rocks, Kelvin asserted the Earth could be about 100 million years old (radioactivity had not yet been discovered, so Kelvin's error can perhaps be excused). Similarly, the sun could be hot from the gravitational potential energy of its formation – but that meant that the sun, too, might only be 100 million years old, and that it was cooling down! That meant that more ice ages might be likely in the future, and it is against this perspective that studies of the other planets proceeded.

Other evidence of the age of the Earth emerged from calculations essentially quantifying the ideas of Halley and Hutton. Geologist and engineer Mellard Reade in 1879 calculated[58] that the dissolved calcium and magnesium salts in British rivers corresponded to a dissolution of the rocks denuding the land by one foot in 12,978 years. Further, scaling this up to the flow of various rivers worldwide, he calculated it would take 25 million years for the sulphate salts to accumulate in the ocean to their present levels. Reade noted that chloride salts would take even longer – 200 million years – establishing this as a lower limit for the age of the Earth.

Other contemplations of the rocks suggested that perhaps conditions had been very different in the deep past. The American geologist T. C. Chamberlin noted[59] in 1898 that limestone rocks easily held more than 60 times the amount of carbon dioxide than was present in the atmosphere. This made changes in the accumulation or removal of carbon dioxide a powerful lever to push the climate, and for this reason he advocated[60] that greenhouse changes were responsible for the ice ages (he played a large role in mapping out glacial deposits in the northern United States, recognizing that there was not one Ice Age, but several episodes of glaciation. He also founded the Journal of Geology in which his glacial speculations appeared).

Chamberlin drew attention a few years later to the possibility that changes in ocean circulation could lead to large climate variations, writing[61] in 1906 '*In an endeavor to find some measure of the rate of the abyssmal circulation, it became clear that the agencies which influence the deep-sea movements in opposite phases were very nearly balanced. From this sprang the suggestion that, if their relative values were changed to the extent implied by geological evidence, there might be a reversal of the direction of the deep-sea circulation and that this might throw light on some of the strange climatic phenomena of the past and give us a new means of forecast of climatic states in the future'*.

Swedish scientist Nils Ekholm reviewed the state of the art of climate science in 1901[62]. He noted that the hot surface of the Earth early in its history would have had a thick, steam atmosphere. Unsurprisingly adopting the perspective of his friend and compatriot Arrhenius, he noted that a sufficient quantity of carbonic acid will produce not only a warm climate, but also a uniform one over the whole Earth. He recognized that the 'carbonic acid trade of nature is carried on with very little capital and a very great exchange'.

Ekholm also introduces the idea that climate change might modulate volcanism. He calculated that if global temperatures dropped by 20 °C, then the crust of the Earth would cool by a comparable amount within some millions of years, whereas the interior of the Earth would remain essentially as hot as before. The crustal cooling would lead to a contraction of some 12km of the circumference of the Earth, inducing 'a great many fissures and subsidences in the regions formerly crumpled, with accompanying volcanic eruptions' to occur. In fact, a century later, the influence of climate on planetary tectonics would be revisited, in the context of Earth's sister planet, Venus.

More useful than this speculation, however, is his analysis of historical Scandinavian temperature records and botanical indicators suggesting warm conditions between 7000 and 10,000 years before the present. He notes that, in fact, variations of the Earth's obliquity can explain the general deterioration of climate since then , calculating the excess in northern polar insolation 9100 years ago, and deducing that summertime temperatures would have been 4 °C or so higher than now. In contrast, 28,000 years earlier, they would have been 7 °C cooler – consistent with the ice sheets being present. Ekholm generously recognizes Croll's priority in developing the astronomically-forced model, although notes that Croll considered principally the changes in the Earth's eccentricity, not its obliquity.

Indirect climate records such as pollen are called 'proxies'. The abundant pollen of a small tundra flower, Dryas octopetala, preserved in Scandinavian bogs about 12,000 years ago, now gives its name to a particularly cold 800 year period in northern hemisphere climate, the Younger Dryas. The Dryas pollen is supplanted by that of trees when forests flourish in warmer conditions.

Ekholm's wide-ranging paper also considers more recent changes, looking at historical records of Baltic sea-ice and (where they existed) temperature records. He concludes that there was 'probably' a change towards a milder, more maritime climate in Northern Europe in the last 1000 years, but recognizes that a cause cannot be determined.

He closes by noting that the cooling of the Earth's interior would result in a decline in volcanism and thus slower replenishment of the CO_2, but that this decline would not be significant for millions of years. On the other hand, he notes that human activity added a tenth of a percent to the atmosphere's CO_2 inventory each year, and that '*if this continues for some thousand years it will undoubtedly cause a very obvious rise of the mean temperature of the Earth*'. But with the calculated obliquity variations pushing Northern Europe towards a new Ice Age in the next 10,000 years, he thought that this global warming by CO_2 might be a good thing, and might be done deliberately.

Meanwhile, Gilbert Walker, a British mathematician working for the Indian Meteorological Department, devised techniques for analyzing time series data[63] and in the early 20th century determined a statistical basis for teleconnections in the Earth climate system. Specifically (helped by a small army of Indian mathematicians), he identified[64] a seesaw anticorrelation between atmospheric pressure over the Indian Ocean and the Pacific (the 'Southern Oscillation' – to distinguish it from another correlation he found, the North Atlantic Oscillation), which also related to temperature and rainfall patterns, and in particular the agriculturally-vital Indian monsoon which had failed causing a famine in 1899. This variation is now known as the El Niño-Southern Oscillation (ENSO), as it also manifests in warmer winter sea surface temperatures off the coast of Peru, where fishermen associated it with the Christmas season, hence El Niño (Spanish for 'The Boy'). During the 'normal' (La Niña) phase, easterly surface winds pile warm water towards Indonesia (such that the sea level is 60cm higher than on the South American coast, where waters are 3-5 °C cooler than normal). These easterly winds form part of a cell, termed the Walker circulation, with warm, wet air rising (with the low pressure Walker noticed) at its western end, and descending at the eastern end. This phenomenon, relying in part on heat stored in the ocean, is one of the principal causes of interannual variation in the terrestrial climate system and adds quasiperiodic 'noise' to climate variables, making the confident detection of secular changes due to increasing greenhouse gases or other factors more difficult. Walker wrote in 1925 *"It is a natural supposition that there should be in weather free oscillations with fixed natural periods, and that these oscillations should persist except when some external disturbance produces discontinuous changes in phase or amplitude"*. Climate might wobble as well as drift.

The notion that carbon dioxide might be a major constituent of the Martian atmosphere was realized around this period by the Anglo-Irish physicist G. Johnstone Stoney (who also invented the word 'electron' to denote the fundamental quantity of electricity). Stoney, who had worked in his undergraduate days around 1850 as an assistant at Rosse's observatory, applied the kinetic theory of gases to planetary atmospheres even in the 1860s, although much of his work was published in the Royal Dublin Society's transactions, where it attracted little international attention. In a more widely-read paper in 1898, he argued[65] that a few molecules in a gas moved at speeds ten or so times the mean square speed (which depends on the temperature of the gas and on its molecular weight) and if that speed was greater than the escape velocity of the planet or moon, then that gas could escape and so should not be found. With that reasoning (more commonly referred to as Jeans escape, after the English astronomer James Jeans who popularized the process in a book a few years later) it was no surprise that the moon had no atmosphere, since its escape velocity was so low. Stoney noted that hydrogen is supplied to Earth's atmosphere by volcanoes, but notes that it should be able to escape to space. While he saw – somewhat anticipating James Lovelock's ideas in the 1960s-1970s – that today, hydrogen would likely be consumed by combustion with oxygen if it accumulated to any degree. He noted that Earth's early atmosphere before oxygen became as abundant as today might allow hydrogen to persist chemically. But he also saw that the light hydrogen molecules would be able to escape to space, so any such hydrogen-rich episode must have been short-lived. And similarly, he noted, water vapor should be able to escape from Mars. The implications were profound : *"Without water, there can be no vegetation upon Mars, at least not such vegetation as we know; and in the absence of vegetation it is not likely that there is much free oxygen. Under these circumstances, the analogy of the Earth suggests that the atmosphere of Mars consists mainly of nitrogen, argon and carbon dioxide."* As indeed it turned out to be. Stoney furthermore suggested that the frosts and fogs seen on Mars were of carbon dioxide, with that heavy gas concentrating in the lowest places, and even that 'extensive displacements of the vapor, consequent upon its distillation towards two poles alternately' might account for the changes in Mars' appearance through the telescope. While Mars atmosphere is generally better-mixed than Stoney perhaps imagined (and argon and nitrogen are only minor constituents), his guess at the composition, and his picture of the carbon dioxide frost cycle, were spot on. But they would not be popular.

Meanwhile, the extent of mixing in the Earth's atmosphere began to be explored. After Glaisher's near-fatal experience in high-altitude ballooning, and indeed the deaths of two of three French balloonists in 1875, exploration of the upper air had stalled until the 1890s with the technical innovation, originating with Georges Besancon and Gustave Hermite in France, of the (uncrewed) sounding balloon. This was a small, lightweight balloon (initially paper, silk or goldbeater's skin - the outer skin of an ox intestine, used in making gold leaf) equipped with instruments that recorded information on paper or metal foil : when the balloon burst, the instrument package descended by parachute and was recovered. Their most successful balloon was named Aerophile. Two challenges were the intense cold and the difficulty of measuring the temperature of the air without contamination from heating by direct sunlight, a much bigger problem in thin air at high altitude (or, 75 years later, on Mars) than on the ground. Teisserenc de Bort in Paris, and Richard Assmann in Berlin launched numerous balloons in the late 1890s and early 1900s; de Bort launched some 236 to 11km altitude, of which 74 reached 14km.

They became confident (Assmann in particular introduced an aspirated thermometer that tackled the sunlight problem, as well as rubber balloons of the modern type) after repeated observations[66] that in fact the progressive decline in temperatures ('lapse rate') with altitude came to a stop, and perhaps even turned over, above an altitude that varied between 8 and 13km. They had discovered, and de Bort named, the stratosphere. And of course one had to come up with a name for the layer beneath it, the troposphere. The French term for the sounding balloons 'sonde' was also adopted internationally, and the higher-tech version that used radio to transmit its readings so it didn't need to be recovered, was called a 'radiosonde' after its development by Pavel Molchanov in Saint Petersberg, Russia in 1930.

Meteorological balloons became an established tool in other

countries too[67], and international conventions arose to synchronize the launch of balloons (in the early 1900s the International Aeronautical Commission had set the first Thursday in each month for such observations) to measure the profile of the atmosphere at many locations simultaneously. Meanwhile, the fundamentals of practical space travel began to be laid out. Konstantin Tsiolkovksy in Russia devised the fundamental equation of rocketry in 1897 and his 1903 'Exploration of Outer Space by Means of Rocket Devices' showed that the ~8 km/s speed change needed to enter orbit could be reached using a multistage rocket fueled by liquid oxygen and liquid hydrogen. Robert Goddard in the USA made the first practical experiments[68] in liquid-fueled rocketry in the 1920s, and suggested that rocket-propelled vehicles would be a useful means of lofting automatic recording instruments to the upper atmosphere and might one day even photograph the moon and planets[69]. Goddard later experimented with gyro-stabilization of rockets - automatic control systems would eventually become a key part of spacecraft and rocket launch vehicles, and in fact the first pilotless aircraft guided by Elmer Sperry's gyroscopes flew in 1917-18, only a few years after the Wright brothers.

The late 19th century saw larger telescopes and improved mapping of Mars[70], but recording by sketches what the human eye sees is a notoriously subjective means of recording data. Wealthy Percival Lowell, inspired by Flammarion's book on Mars and Schiaparelli's report of 'canali', founded a world-class observatory in Flagstaff, Arizona, at about 2000m elevation, just south of the Grand Canyon. Whereas modern observatories are typically on higher mountains still, in fact, for visual astronomy this elevation is actually optimal[71] due to the physiological effects of high altitude on the observer.

Lowell convinced himself he saw a network of dark lines on

Mars, and conjured a popular vision of a dying world inhabited by an intelligent civilization, conveying the last dregs of its water from the polar caps to lower latitudes. Yet others were unpersuaded – Maunder in England, and even Andrew Douglass, who had scouted locations for Lowell's observatory and later ran it on Lowell's behalf, found that the eye tends to join dots seen on a small disk through a telescope, and suspicions became widespread that in fact the canals were illusory (see figure 2.7). Douglass was fired, and moved to Tucson in southern Arizona where he founded not only an observatory, but also a laboratory to study tree rings as an indicator of past climate and solar activity. (Ironically there is today a canal, the Central Arizona Project, which brings water from the Colorado river west of Flagstaff down to parched Tucson – not perhaps unlike Lowell's vision of Mars.)

Another astronomer, in Catalonia, was also gifted with a good telescope, good atmospheric conditions, and eyesight. At his Fabra Observatory[72], overlooking Barcelona, Jose Comas Sola could tell with his 37-cm binocular telescope[73] that Mars had no canals. He was also able to tell that tiny Titan, whose disk was just visible in the telescope, was like Neptune in that the edge of the disk was subtly darkened, unlike the sharp edge of the Moon. This, he correctly interpreted in 1907, meant that Titan had an atmosphere.

Lowell published an estimate of the climate of Mars[74], using the fourth-power law. The trick here is that neither the solar constant, nor the effects of the Earth or Mars' atmosphere, nor the reflectivity of the Earth were known. Lowell was aware (and quotes a calculation by Moulton) that if the Earth and Mars have the same reflectivity, then without atmospheric effects, the absolute temperature of Mars should be the fourth root of (4/9) times that of the Earth, since Mars is (3/2) times further away, and thus gets (9/4) times less sunlight. That means about -30 °C.

He recognized that the Martian atmosphere is thinner than ours, but has more carbon dioxide, so he considers these factors to balance out. But he argued that the Earth and its cloudy atmosphere had an albedo, or reflectivity, of some 75%, whereas that of dusty Mars is only a third of that value. This factor of 3 in absorption completely compensates for Mars' further distance from the sun, such that Mars ends up being about the same temperature as the Earth.

It is difficult not to see this calculation as a case of Lowell pushing the numbers to get the answer he was looking for, and indeed a response followed swiftly. Poynting in England quickly dissected[75] Lowell's calculations and noted that white cardboard has a reflectivity of nearly 75% but looks much brighter than the sky, so the foundation of Lowell's calculation was likely way off. Lowell was prolific in the popular media as well as in the learned journals, and among his books, 'The Canals of Mars' prompted a fierce book-length rebuttal[76] by Alfred Russel Wallace, the co-discoverer with Darwin of the Theory of Evolution by Natural Selection. Wallace notes that on Earth at least, temperatures fall with elevation, so that when the atmosphere above becomes thinner, the reduction in blanketing effect is a stronger effect than the reduction in obscuration of sunlight. Regardless of the apparent delusions of Lowell about the canals, and the deluge of scientific criticism his work drew, Lowell did much to popularize astronomy and contemplation of the habitability of the planets, and the Observatory he founded remains a respected and productive institution today.

Arrhenius[77] also judged Mars to be colder than Lowell had advocated, suggesting an average temperature of -40 °C, or -10 °C at the equator. But he suggested that surface liquids could still exist due to the antifreeze effect of calcium and magnesium salts. He recognized that there must be but little water vapor in the cold atmosphere, with clouds being therefore rare, but imagined (rather presciently) that in summer, hoar-frost on the polar cap would evaporate and as the relatively humid air swept over hygroscopic salts, these would absorb the water and form local wetness, a process that would hit the headlines a century later. Ironically, Arrhenius failed to consider a greenhouse effect on Venus, calculating its average temperature to be 47 °C, torrid but not barren. Swirled in clouds, and with the near-surface atmosphere holding '*three times the humidity of the Congo*' he imagined Venus to be dripping wet, covered with luxuriant but primitive vegetation. Interestingly he suggests that despite the clouds and moisture, precipitation might not be higher than on Earth, the thick clouds suppressing air currents.

In this era, the behavior of gas exchanging heat by radiation in a gravity field began to be understood, stimulated in part by the discovery of the stratosphere[78]. Much of this work appeared first in the German language (by Austrian, Swiss and German scientists) and much of it in the astrophysical literature. Beginning with work on the structure of stars by Karl Swarzschild in 1906, other analyses exploring the fundamentals of how the Earth's stratosphere came about soon followed by E. Gold and by W. Humphreys (in the USA) in 1909 : Humphreys rather presciently suggested that absorption by ozone might be responsible for warming the Earth's stratosphere. Robert Emden's[79] work *Gaskugeln: Anwendungen der mechanischen Wärmetheorie auf kosmologische und meteorologische Probleme* (Gas balls: Applications of the mechanical heat theory to cosmological and meteorological problems) sums up the dual applications of this radiative equilibrium theory.

These early works assumed 'semi-gray' atmospheres, where the absorption at different wavelengths is the same[80]. Soon, however, some inconsistencies had to be resolved : Hugo Hergesell in Germany noted that the colder upper troposphere couldn't hold enough water vapor for the assumed variation of transparency to hold. Edward Milne in Cambridge and R. Mugge explored how the exchange of radiation should vary with latitude, the latter also calculating more generally what Croll had started to consider, namely how much heat was transported by the atmosphere as a function of latitude.

Some inconsistencies remained, however, and were finally cleared up in a rather important set of theoretical papers by George Simpson[81], a prominent meteorologist who in fact had accompanied Scott's expedition to the Antarctic. In computing heat fluxes, assuming that water vapor was the only significant absorber (a prejudice that may have influenced his reception of Callendar's suggestion of a strengthening carbon dioxide greenhouse a decade later) he found the interesting result that there was in fact an upper limit to how much radiation the atmosphere could export to space. This paradox was the kernel of the 'runaway greenhouse' idea that would eventually prove important in understanding Venus.

While the semi-grey methods used by these workers would still be used extensively in the study of climate on other planets, it became recognized at this point that they really were not adequate to explain what had become known about the Earth (in fact Gold had already noted that there had to be part of the spectrum where the Earth's surface could radiate unimpeded). But the calculations began to challenge the ability of scientists to compute by hand.

Meanwhile, the utility of physics for understanding, and more particularly predicting, the short-term weather on Earth began to be recognized. Vilhelm Bjerknes in Sweden in 1907 suggested that weather might be predictable, not from the time-honored expedient of using the weather as a model of itself (i.e. look for when the weather pattern was like today in the past, and then expect tomorrow what happened the day afterwards – an approach dating back to the Greek scientist Theophrastus, and systematically employed by Fitzroy and others) but rather by applying physical laws. In fact, in 1901 the prominent American meteorologist Cleveland Abbe had argued in a long paper[82] in the Monthly Weather Review (a journal he founded in 1872) that long-range weather forecasts should use equations of conservation of energy and momentum. *"By thus considering the land and water hemispheres of our globe as the thermal and frictional disturbers of the phenomena that would otherwise pertain to a uniform surface, rapidly rotating... we shall arrive at the desired result sooner and better by the study of the mechanics of the atmosphere than by the search for elusive empiric periodicities"*.

The outbreak of the First World War interrupted many scientific investigations. Fortunately, the Serbian mathematician, Milutin Milanković was interned only briefly[83], and in Budapest contemplated the current climate of inner planets of the solar system. In 1916 he published a paper entitled "Investigation of the climate of the planet Mars". Milanković calculated that the average temperature in the lower layers the atmosphere on Mars is −45 °C and the average surface temperature is −17 °C. Also, he concluded that: *"This large temperature difference between the ground and lower layers of the atmosphere is not unexpected. Great transparency for solar radiation makes the climate of Mars very similar to high-altitude climate of our Earth."* In any case, Milanković theoretically demonstrated that Mars has an extremely harsh climate.

Milanković went on to evaluate with more mathematical rigor[84] the astronomically-forced changes in the Earth's climate that led to repeated glaciations. While sometimes referred to by his name, these variations – which we now know also to occur on Mars and Titan – were essentially an elaboration on the mechanisms laid out by Croll, and are often termed more appropriately 'Croll-Milanković cycles'.

Elsewhere in the European conflict, Lewis Fry Richardson[85], a bespectacled British ambulanceman dealing with the horrific casualties in France attempted - by hand, taking months - to calculate a six-hour change in weather conditions (starting from 7am on 20 May, 1910 – a day when coordinated balloon measurements had been made across Europe). His result was inaccurate, due to poor quality of his input data, but the mathematical framework was sound, and forms the foundation of modern numerical meteorology. This framework – essentially distilling the state of the atmosphere into a table of numbers by dividing the world up with a rectangular grid and applying equations to process the numbers - was outlined in a book, the manuscript for which was temporarily lost under a heap of coal somewhere behind the front lines where it was sent for safety. Happily it was recovered, and eventually published in 1922.

Richardson envisioned that these calculations would be made by people, working like an orchestra. A conductor would be needed to coordinate the calculations, making sure that one part of the grid did not outpace the others. Since we take modern forecasts for granted, it is worth repeating a little of the preface of 'Weather Prediction by Numerical Process',

' in this book, a scheme of weather prediction, which resembles the process by which the Nautical Almanac is produced, in so far as it is rounded upon the differential equations and not upon the partial recurrence of phenomena in their ensemble........ The scheme is complicated because the atmosphere is complicated......Perhaps someday in the dim future it will be possible to advance the computations faster than the weather advances, and at a cost less than the saving to mankind due to the information gained. But that is a dream.'

Richardson, a pacifist, would perhaps be horrified to learn that another World War would see his dream realized, with the development of machines able to perform the needed calculations fast enough, as we describe in the next chapter. Indeed, the computational appetite of weather prediction and climate simulation has advanced to the point where massively-parallel computer clusters are now used, and the challenges of orchestrating distributed calculations that Richardson imagined are important issues today.

4

FEELING THE HEAT

1920 - 1960

After World War I, bigger telescopes and more sensitive thermopiles opened the way to measuring the temperature of the planets directly, by their infrared emission. Edison Pettit and Seth Nicholson[86] used the 100-inch Hooker Telescope on Mount Wilson in 1923 to measure the center of the Mars disk (i.e. roughly at noon) as just above freezing, while the polar cap was at -70 °C. They also found the variation of infrared emission from the limb to the center was like that of the moon, with no displacement of the warm peak due to rotation – this meant that, like the moon, Mars' surface warmed and cooled quickly. Photographic methods were also improving, allowing a record of the planet's appearance that did not rely on astronomers' sometimes-unreliable vision and memory.

It was still thought at this time that nitrogen and argon might be the principal constituents of the Martian atmosphere, and while likely thinner than Earth's (as Herschel's visual observation of stellar occultation had indicated) astronomers struggled to measure just how thin. Donald Menzel at the Lowell Observatory reported on the basis of photometric measurements in 1926 an upper limit might be 66 mbar, while B. Lyot in Paris based an upper limit of 24 mbar on measures of the polarization of light. These challenges, and perhaps some wishful thinking, allowed scientists to continue to labor under the illusion that the atmosphere might be perhaps as high as 85 mbar, as advocated by Gerard de Vacouleurs (also in France) from the 1930s, in part by estimating how the contrast of Martian surface features varied as the planet's rotation brought them to the edge of the disk[87]. All these methods, however, methods neglected the role of scattering by dust suspended in the atmosphere, causing the pressure to be overestimated.

Dust became a prominent topic in the 1930s in the USA, as a poor choice of farming methods rendered the ground susceptible to erosion, especially in Texas, Kansas and Oklahoma. Cultivation that had been tolerable in the relatively wet 1920s failed as drought blighted crops and strong winds swept soil up into massive dust storms. The agricultural and economic upheaval caused by these impacts – sometimes blotting out the sun and bringing visibility down to a meter – intensified the suffering of the Great Depression, and precipitated the migration of millions from the plains states.

Meanwhile, in Egypt, the first systematic survey[88] of dust devil activity was made by W. D. Flower, a scientist working for the British Meteorological Office (at the time, part of His Majesty's Air Ministry – as in the USA, weather services began as branches of the military). He documented their numbers, sizes and behaviour in several outposts of the Empire: Jordan, Iraq, Palestine and the Sudan as well as Egypt. Flower's careful records have been useful even three quarters of a century later, to identify characteristics that have been compared with dust devils on Mars[89].

In 1928, Richard Courant, a German mathematician, outlined[90] a criterion for the stable convergence of partial differential equations by finite difference methods. While that may sound abstract, it represents a law on numerical climate simulation as inconvenient yet binding as the finite speed of light is on space travel. Essentially the Courant (or Courant-Friedrichs-Lewy) criterion states that when simulating weather or climate dynamics in the way Richardson had proposed, the time step over which the calculations are marched must be small enough. The failure of Richardson's example attempt was in part due to over-reaching, violating this criterion, which also states that the smaller the grid cell on which the calculations are made, the smaller the time step must be – multiplying the number of calculations even more as a finer grid is proposed. Numerical weather prediction would have to wait.

In the remote northwestern parts of the USA, some indirect but profound impacts of the Ice Ages, and in particular the process of deglaciation, began to be recognized. Geologist J. Harlen Bretz argued[91] that the massive valleys and waterfalls carved into hard basalt rock in what he called the channeled scablands of Washington state could only be made by an immense flow of water. With J. T. Pardee, he recognized that massive Lake Missoula, previously dammed by the ice sheets, had drained suddenly as the ice receded. The deep canyons, falls, submarine dunes and giant strewn boulders attested to massive flow – a megaflood. Such catastrophism took some decades to be widely embraced. This region, and the event that shaped it, became a prototype for the outflow channels later to be identified on Mars[92].

Yet more revolutionary than the idea of glaciation and catastrophic floods, perhaps, was the idea that the Earth's continents have moved relative to one another. In 1912, Alfred Wegener, a German geophysicist (with strong interests in meteorology - Wegener and his brother Kurt set a world record for ballooning in 1906 on a scientific flight lasting 52 hours) proposed continental drift. In 1924, Wegener and his father-in-law, Russian meteorologist Wladimir Koppen, who in turn had defined a prominent system of climate classification in the

1880s, wrote a book[93] which lent support to (and drew from) Milanković's work, and also embraced continental drift to explain how places such as Svalbard[94] at 78°N could have coal deposits, which required lush vegetation in the deep past. Once continental drift is embraced, the story is simple – Svalbard wasn't in the arctic when the coal-forming vegetation was laid down. At the time, the best evidence in support of continental drift was the fossil record and the match of the coastline shapes of Africa and South America, which surely strikes every child confronted with a map : it was not until magnetized stripes, symmetric about the mid-Atlantic ridge, were measured in the 1950s demonstrated seafloor spreading that the concept was universally accepted.

Coal, and human combustion of it, entered the climate picture in 1938, when Guy Stewart Callendar, a British steam engineer and amateur meteorologist, proposed[95] that the atmospheric carbon dioxide concentration had measurably increased, from around 290 parts per million (ppm) before 1900 to about 300 in the 1930s, although the uncertainties on these measurements were not well-determined. He had also carefully compiled temperature records from around the globe, and found a general increase. And then Callendar joined the dots – he not only argued that the carbon dioxide increase was consistent with fossil fuel combustion (and that the extra CO_2 had not been taken up completely by the oceans), but also presented calculations showing that the CO_2 increase could explain the apparent temperature increase.

Callendar's unfunded but prescient work received only modest attention at the time, perhaps in part due to a dismissive response from the prominent Sir George Simpson, who praised the amount of work Callendar had put into the paper but observed (my underlining) '*that it was not sufficiently realized by <u>non-meteorologists</u> who came for the first time to help the Society in* *its study, that it was impossible to solve the problem of the temperature distribution in the atmosphere by working out the radiation'.* While Simpson's point that convection plays a role in atmospheric structure was correct, his condescension in referring to Callendar as a newcomer and 'non-meteorologist' seems brutal.

In fact, Callendar's usual attribution as a 'steam engineer', which perhaps conjures an image of a grease-blackened brute with a large wrench, fails to really recognize that his career was all about precision measurement. He had presented scientific measurements on the thermodynamic properties of high-pressure steam at international conferences, so while his work on climate was performed in his spare time, his analysis was by no means amateurish.

Simpson dismissed Callendar's empirical observation of warming as a spurious change : *'the rise in temperature was probably only one phase of one of the peculiar variations which all meteorological elements experienced'.* Indeed, global temperatures underwent a dip in the 1960s which meant Callendar's observations and predictions received relatively little attention. However, recent assessments of the evolution of surface temperatures and CO_2 concentration show that Callendar's predictions were remarkably accurate. [96]

About the same time, the idea of a carbon dioxide greenhouse on Venus was advanced by Rupert Wildt, a German astronomer in the USA. Drawing on spectroscopic observations of CO_2 in the Venusian atmosphere (by Adams and Dunham in 1932, using special new film emulsions sensitive in the near-infrared[97]) and laboratory data on CO_2 opacity he made the rudimentary calculation that Venus was "probably warmer than the boiling point of water".[98] The amount of CO_2 estimated was only, however, that amount near and above the cloud-tops, so the strength of the greenhouse, and thus the surface temperature, were considerably underestimated.

1938/9 also marked the determination by Hans Bethe of the nuclear fusion reactions that power the sun. Arthur Eddington in 1920 had suggested that fusion of hydrogen is what produced helium, and energy, in the sun. George Gamow in 1928 had explained with quantum mechanics how the hydrogen nuclei might furtively tunnel through the barrier to join together. With this understanding of the basic process fueling the stars – physics that was simply unknown to Kelvin and Helmholz when they confronted the ages of the sun and the Earth at the turn of the century -   and models of stellar structure starting with Schwarzschild's, the evolution of stars could now be predicted.

World War 2 intervened at this point, bringing a new world order, as well as a vast demand for meteorological products to support military operations. The needs for aviation[99] were obvious, and meteorology even attained a planetary scale in its application. Although the discovery of Earth's jetstreams is often credited to the high level of Allied bomber activity which yielded consistent observations, in fact the northern hemisphere jet was discovered by Japanese[100] meteorologist Wasaburo Ooishi in the 1920s. Launching small, passive balloons (pilot balloons or pibals) from Tateno, Japan and tracking them by theodolite (Mt. Fuji was 160km away, and served as a useful target to assess atmospheric visibility before launch) he measured winter westerly winds at up to 70 m/s at 10km altitude. In fact, these winds were exploited during the war to launch some nine thousand incendiary-carrying balloons[101] against the USA in 1944-1945 - the first intercontinental weapon. About 300 are known to have reached the Americas, but caused little damage[102].

Another wartime development of note was the development of

methods by Sverdrup and Munk[103] to predict of wave heights at sea, to determine whether amphibious landings would be feasible. Despite the best efforts of mathematicians for over a century, the fundamental prediction of wave growth in response to sea-surface wind stress is a challenging area, and Sverdrup and Munk's semiempirical approach struck a useful balance between theoretical rigor and empirical expedience : 70 years later I was to largely follow their methodology in predicting waves[104] on Titan's sea Ligeia Mare for proposed capsule to float there.

In 1944 the Dutch-born astronomer Gerard Kuiper made spectroscopic observations of the planets and Titan, and found remarkably that its spectrum showed dark bands which matched laboratory measurements of absorption bands of methane. Titan had a significant atmosphere, making it unlike any other satellite (and promoting it to being a subject of this book.) While Comas Sola's visual observation had been correct, it was not reproducible, whereas a photographic spectrum was much more solid evidence.[105] Kuiper also showed[106] with spectra he acquired in 1947 that carbon dioxide was present in the Martian atmosphere (nitrogen and argon were suspected by analogy with Earth, as noted by Stoney, but these simple gases do not have easily-detectable spectroscopic signatures and so were not actually detected until spacecraft measurements were made much later).

World War 2 stimulated many technological developments that directly or indirectly influenced the study of planetary climate. Rocketry, notably the liquid-fueled and gyro-stabilized German V2, would pave the way for the vehicles that would make planetary exploration possible. The war, and the Cold War that followed, saw improvements in electronics of all kinds that eventually filtered into astronomy. Detectors of infrared radiation intended for heat-seeking missiles, together with improved models of the propagation of such radiation through the atmosphere made major progress, especially in the 1950s and 1960s both for improving telescopic measurements of planetary temperature, and in modeling the effect of gases on climate. Similarly, the development of radar[107] provided a platform for major advances in radio astronomy, which opened a new window into the study of planetary temperatures.

When turned to our nearest neighbor planet, radio methods indicated Venusian temperatures of many hundreds of degrees. The first reliable estimate[108] was made using observations in 1956 with a 50-foot radio dish at the Naval Research Laboratory in Washington, D.C. (in fact the radio telescope was calibrated with a radio transmitter mounted at the top of the Washington Monument!)

This result was a puzzle, as infrared measurements by Sinton and Strong[109] in 1960 showed that the day and night temperatures in the atmosphere were about 240K, in agreement with earlier work by Pettit and Nicholson[110] (who actually took their measurements in the 1920s). The infrared measurements were probing the cloud tops, but what was the radio seeing? Jupiter was a strong radio source, but not because it was hot. Despite this uncertainty, Mayer and colleagues estimated the radio brightness of Venus at their 3 cm wavelength to be about 600 K, not too far from the right value. This very high temperature was a great puzzle, and would not be resolved without traveling to Venus itself.

Meanwhile wartime advances in nuclear physics paved the way for radiocarbon methods and the isotopic analyses which would ultimately be the key to reconstructing Earth's climate. But it was the development of electronic computers, stimulated in part by the need to compute ballistic trajectories for gunnery[111] that finally permitted the first practical experiments in weather prediction by numerical expressions of physical laws as envisioned by Abbe, Bjerknes and Richardson, rather than the educated guesswork by graphical methods of human forecasters.

Influential individuals in these steps included the Swede Carl-Gustav Rossby[112] and the computer scientist John von Neumann, but the person most identified with the 'first' electronic computation for weather prediction is Jule Charney, hired by Neumann at Princeton, who broke down the atmospheric dynamics equations into something tractable for computation by the ENIAC[113] computer. These first steps in weather computation[114] were regional, not global, with the initial conditions set by meteorological observations rather than first principles, and the output propagated only 24 or 48 hours ahead. Although numerical climate prediction is in principle the same problem, the practical details of the approach are different. In weather prediction, the assimilation of vast amounts of observational data, and the forward propagation of the system by approximations that are 'good enough' but streamlined to work at very small scales (high resolution) is the overall approach, whereas for climate prediction, it is more important to get the physics right, albeit at lower resolution (bigger 'boxes'). The first global simulation, by Norman Philips soon thereafter, found just this challenge with the physics – the abstract to his paper[115] concludes *'Truncation errors eventually put an end to the forecast by producing a large fictitious increase in energy.'*. Ultimately, and with enough model refinement and computer power, the two modeling philosophies converge, but the emphasis in the 1950s at least was on weather forecasting and dynamics, rather than climate.

In the 1950s some graphic insights into heat transfer in a rotating fluid were developed with laboratory experiments, first by Dave Fultz and Herbert Riehl in Chicago and later by Raymond Hide in Cambridge[116]. These so-called 'dishpan experiments', with cold centers and heated rims generated long waves, vortices, and meandering jetstreams. While these analog experiments were not predictively useful in the same way computer models were, they helped thinking about the dependence of the circulation regime on the atmospheric and oceanic boundary conditions like rotation rate (which would be useful in understanding the differences between different planets) and served to give tractable benchmarks for computers to reproduce.

Some basic aspects of the ocean circulation began to be understood around this time, with Henry Stommel of the Woods Hole Oceanographic Institution developing some basic ideas on how wind stress acting on the ocean surface of the rotating earth would form Western Boundary Currents (like the Gulf Stream and the Kurishiro current at Japan) on the western margins of the midlatitude oceans. He further proposed simple models of the way salinity and temperature drive the deep ocean circulation, with cold salty water at high latitudes (salty because as sea ice forms, the salt stays in the liquid) sinking into the abyssal depths whence it rolls equatorward. He constructed a simple box model of this 'thermohaline' circulation[117], and found that it had two steady states, wondering *"whether other, quite different, states of flow are permissable in the ocean or some estuaries, and if such a system might jump into one of them with a sufficient perturbation. If so, the system is inherently fraught with possibilities for speculation about climate change".*

The post-war years saw the first polar ice coring[118] – in 1949-52 British-Norwegian-Swedish, USA and French teams drilled ~100m-long cores in Antarctica, Alaska and Greenland respectively. The idea that annual cycles might be discerned in the firn (packed snow) and ice layers, forming a nice consistent record of the past climate was first demonstrated in a 15m pit hand-dug by Ernst Sorge in 1930, on Wegener's expedition to Greenland. Drilling, and especially sample handling and analysis instrumentation, progressively improved. One stimulus was the International Geophysical Year (IGY, 1957) - and US-Danish projects (Greenland being Danish territory) could drill over a kilometer down to bedrock by the early 1970s, claiming "*One thousand centuries of climate record from Camp Century on the Greenland Ice Sheet*". These records provide our best understanding of climate over the last million years or so – not only does the isotopic composition of the water shed light on temperatures of the past, but tiny bubbles of gas trapped in the ice allow variations in the greenhouse effect to be quantified, and layers with sulphate acidity attest to volcanic eruptions.

In 1952 in Ireland, the Estonian astronomer Öpik[119], accounted for heat transport in a simple model of the Earth's climate (which he was attempting to use, in conjunction with the record of glaciation, to demonstrate his rather bizzare theories on fluctuations of solar luminosity.) While these papers[120] are - probably justifiably - virtually unknown, among them is a notable attempt at a global climate model (specifically, a one-dimensional zonal energy balance model) that predates the more famous incarnations by Budyko and Sellers by two decades.

Other scientists also favored solar variations and/or volcanic aerosols as the cause for climate change. Harry Wexler, a prominent US meteorologist[121] in 1956 advocated these effects[122], yet barely mentions carbon dioxide. But after a decade or two of inattention, the hypotheses advanced by Callendar received renewed examination. First, Gilbert Plass, a US physicist[123], examined infrared absorptions by carbon dioxide more carefully (using a new-fangled electronic computer to analyze the hundreds of thousands of spectral lines), and largely retired the objection that the CO_2 and water absorption bands overlapped[124], such that the atmospheric absorption at a given wavelength was not already 'saturated'. Thus, adding more CO_2

beyond what was already there would cause warming after all. Second, radiocarbon dating had been devised by 1949, and could be used to assess the amount of carbon being added to the atmosphere. (Basically, a tiny but near-steady fraction of radioactive carbon-14 atoms results from cosmic rays interacting with nitrogen; when this carbon is locked up in a plant, the fraction of carbon-14 slowly declines, with a half-life of about 7000 years. But fossil fuels are very old, all their carbon-14 has decayed, so adding CO_2 from the combustion of fossil fuels dilutes the radiocarbon in the atmosphere.) Radiocarbon analyses[125] enabled Hans Suess to deduce that carbon was indeed accumulating in the atmosphere, although with Roger Revelle, the initial estimates suggested that the oceans might be absorbing much of it. More refined later studies showed this was not the case, however, but those studies were in part stimulated by Suess and Revelle's work – the evolving terrestrial greenhouse was becoming a respectable topic of study.

That, and the healthy scientific support stimulated by the International Geophysical Year in 1957, enabled Charles Keeling to begin a systematic series of carbon dioxide measurements in locations that would be unperturbed by local industrial or agricultural effects, notably on the summit of Mauna Loa in Hawaii. Within just a couple of years[126], his data showed a nice seasonal cycle of a few parts per million (as vegetation, of which most is in the northern hemisphere, took up and released carbon dioxide over the course of a year) but also that the cycle didn't quite repeat – the peak value in one year was one or two ppm higher than the previous one. Although sometimes challenged to maintain funding for the program, Keeling managed to sustain it, yielding what has become an icon of climate change, the Keeling Curve. The plot shows an annual cycle as the boreal forests inhale carbon from the atmosphere in northern spring (rather less land surface covered by trees is present in the southern hemisphere, so the net effect is an annual, rather than biannual cycle), but year by year a steady creep upwards in CO_2 levels occurs. When Keeling's measurements began in 1958, the concentration was 315 ppm ; when Keeling died in 2005 it was 380, and recently it passed the 400 mark.

So much for the Earth. It was known that there was carbon dioxide on Mars and Venus, and the discovery of the amounts will be discussed in the next chapter. But how much water vapor? Water after all is a strong greenhouse gas and is ultimately the determinant of habitability. In principle, spectroscopy would be able to give the answer, but the challenge is to measure the absorption of spectral lines or bands in the remote atmosphere, through the Earth's atmosphere which is itself drenched in the same vapor. Astronomers on both sides of the Atlantic tackled this challenge in two different ways. Myron Spinrad[127] eventually had the most robust success, using a very high resolution spectrograph on Mount Wilson in 1963, in a 4-hour exposure with special ammoniated film, taken not when Mars was closest, but when it was at quadrature – i.e. at its widest angle from the sun. This meant that Mars was moving towards Earth at 15 km/s, enough to Doppler-shift the narrow Mars water lines such that they did not overlap with the terrestrial ones (a technique suggested originally by Lowell, but difficult to implement with the instruments available at the time). The measurement was robust because several different lines were measured – and they indicated that only a few precipitable microns of water vapor existed in the Mars atmosphere, a tiny amount.

In Europe, Adouin Dollfus of the Paris Observatory went to great lengths, or at least dizzying heights, to answer the same question, by getting above as much of the terrestrial atmosphere as he could. After some initial experiments[128] in 1954-1957 to 7 km in a conventional balloon with an open wicker basket (to observe the convective granulation of the sun, and with a spectrometer to study the sky brightness), Dollfus and his aeronaut father[129] developed a remarkable[130] balloon-mounted telescope with a special photometer using a birefringent crystal to select light in and out of a water band, with the telescope operated from inside a pressurized aluminium gondola. Dollfus took the system to an altitude of 13.5km (also in 1959, breaking French ballooning records) suspended, almost comically, beneath a string of 104 weather balloons, which were easier and cheaper to obtain than a single large purpose-built balloon. It is rumored that controlled deflation to descend was effected with a small rifle.

In the end, these quixotic attempts were too much at the mercy of the weather, both from the balloon dynamics point of view, and from variations in the water vapor content of the Earth's atmosphere. Dollfus ended up getting better results from his photometer when it was helicoptered to the top of the Jungfraujoch in the Swiss Alps in 1963 to observe Venus as it was in the same part of the sky as (indeed, was occulted by) the Moon, so that the Moon could act as a calibration reference, peering through the same amount of the Earth's atmosphere as Venus, but without its own water absorption.

In November 1957, with the launch of Sputnik-1, the space age began. Within two years satellite measurements of the Earth's climate from space were attempted. The February 18, 1959 headline in the New York Times was 'Vanguard fires Satellite into Orbit to Scan Weather' – this battery-powered Vanguard 2 satellite was followed up by the solar-powered Explorer 7 in October, which yielded better measurements allowing Verner Suomi to quantify the outgoing thermal emission and cloud cover[131]. The potential for orbital observations to improve weather forecasting was demonstrated by the satellite TIROS-1, which beamed back the first TV images of Earth in 1960, showing the evolution of storms from above. But spacecraft would not remain earth-bound for long – within only five years, we would begin to study the other planets up-close, with the first measurements at Venus in 1962, by a spacecraft whose name evokes many of the early explorations described in this book – Mariner 2.

Feeling the Heat

5

CONCLUSION

The cut and thrust of the scientific process is such that progress is uneven : new ideas can be ignored, or even suppressed, for decades. But the long view gives a rosier picture, as eventually the scientific process winnows ideas that are useful from those that are less useful. Aristotle's world-view of four elements sufficed to explain much of what he saw, phlogiston had a certain consistency with observed chemical transformations, but in time the more modern paradigm of elements proved more effective at explaining how things worked, and at predicting what might become.

And so it has been with planetary climate. The subtleties of absorption of heat and light by the atmosphere, and how conditions on Earth have changed through time, have been progressively unravelled, hand in glove with the determination of the environment on other planetary bodies.

The blanketing effect of planetary atmospheres was recognized centuries ago, but ingenious experiments and challenging mountain and balloon explorations were needed to disentangle the effects of heat and light in the air and quantify these ideas. Ever-better models, in part inspired by trying to understand the structure of stars, have helped predict the Earth's weather and climate, and to explain what we see at Mars – a cold, dry climate, much as was anticipated from the end of the 19th century – and at Venus, where the thick, cloudy atmosphere exerts a truly profound influence on its surface conditions, far more hostile than had been expected.

I have been impressed with the enterprising boldness of many researchers in climate, in climbing mountains, or floating above in balloons, or voyaging to far-off-places, just to obtain measurements. Perhaps this is to be expected – to understand the working of planets is an intellectual exploration, not unreasonably mirrored by adventures in the real world.

It is striking to me that many of the players in the tale of planetary climate have had other roles – attempting to divine the cause and nature of climate has often been just part of a broader portfolio of endeavours. Indeed, while planetary exploration and climate studies are now firmly 'big science' many significant advances in climate seem have been made by amateurs like Croll and Callendar.

That we should love to explore the question of planetary climate is hardly a surprise, it is one that speaks to a deep and innate human curiosity, not only about our immediate prosperity and safety, but our origins and future, and whether we are alone in the universe.

ABOUT THE AUTHOR

Dr. Ralph Lorenz is a planetary scientist and aerospace engineer, originally from Scotland. He worked for the European Space Agency on the Huygens probe to Titan and has been involved in many NASA and international space projects, including Cassini, Mars Polar Lander and JAXA's Akatsuki Venus Climate Orbiter, and studies of planetary balloons, penetrators, quadcopters and even submarines.

His interest in planetary climate arose from anticipating the findings of the Mars Polar Lander and Cassini-Huygens missions, most especially in exploring the Greenhouse effect and equator-to-pole temperature variations on Titan.

He enjoys visiting exotic locations on Earth, from the Arabian desert and Alaska to Vanuatu and New Zealand, to learn about processes on other worlds, notably dust devils, sand dunes and volcanos. He enjoys writing, perhaps something to do with attending King Edward VI Grammar School in Stratford-upon-Avon, the same school as William Shakespeare. He has a B.Eng. in Aerospace Systems Engineering from Southampton University, and a Ph.D. in Physics from the University of Kent in the UK. He spent 12 years at the Lunar and Planetary Laboratory at the University of Arizona, and is now on the Principal Professional Staff in the Space Exploration Sector at the Johns Hopkins Applied Physics Lab in Maryland, USA.

Notes

74

NOTES AND REFERENCES

[1] For a modern version of this measurement, see M. Beech (2010) Atmospheric Height by Twilight's Glow, Journal of the Royal Astronomical Society of Canada, 147-148

[2] I'd never learned of Shen Kuo until I started writing this book. Wikipedia, while not infallible, is an amazingly useful thing. Finding such things is one of the main joys of writing.

[3] Although Galileo is sometimes credited with the invention of the telescope, it was likely invented in the Netherlands in 1608, but quickly spread. Thomas Harriot in England observed the sky with it in 1609. Galileo was an early and enthusiastic adopter of the new invention, and quickly built his own and improved the design, and importantly, wrote down his findings.

[4] G. Tierie, Cornelis Drebbel (1572-1633) Ph.D. Thesis, University of Leiden, 1932. (available online)

[5] Huygens, C. (1698), The Celestial Worlds Discover'd: Or, Conjectures Concerning the Inhabitants, Plants and Productions of the Worlds in the Planets, Timothy Childe, London, U. K.

[6] Galileo had suggested the arrangement of the Jovian moons could be used as a sort of universal clock (for terrestrial, rather than Jovian, sailors to calculate longitude) but this relied on telescopic observations from a heaving ship, which had severe challenges. Halley confronted the problem in various ways, considering occultations of stars by the Moon, or the use of the dip of the Earth's magnetic field. Ultimately it was the development in 1761 of the marine chronometer, a clock that maintained sufficiently accurate time despite temperature changes and the ship's motion, that allowed practical navigation.

[7] Wilkins, J. A Discovery of a New World, or A Discourse tending to prove, that 'tis Probable there may be another Habitable World in the Moon, with a discourse on the probability of a passage thither, J. Gillibrand, London, England. 1684

[8] No human reached 20 miles (32 km) until the 1950s (38.5 km in the Bell X-2 rocket plane in 1956, thirteen years before Apollo 11). Although ballooning began long before heavier-than-air aviation, the 20 mile threshold was almost breached (31 km) in the StratoLab V balloon in 1961.

[9] Halley, E. (1693) A Discourse concerning the Proportional Heat of the Sun in all Latitudes, with the Method of Collecting the Same, as It was Read before the Royal Society in One of their Late Meetings', Philosphical Transactions of the Royal Society, 17, 878-885. The result is expressed in rather arcane terms : ' *to give the proportional degree of Heat or the sum of all the Sines of the Sun's Altitude while he is above the Horizon in any oblique sphere, by reducing it to the finding of the Curve surface of a cylindrick Hoof, or of a given part thereof*.

[10] Halley, E. (1686) An Estimate of the Quantity of Vapour Raised out of the Sea by the Warmth of the Sun; Derived from an Experiment Shown before the Royal Society, at One of Their Late Meetings, Phil. Trans. 1686-1692 16, 366-370. doi: 10.1098/rstl.1686.0067

[11] I made similar sweeping approximations myself in performing essentially the same calculation for methane evaporation from Titan's seas. I think Halley would approve. Lorenz, R. D., 2014. The Flushing of Ligeia : Composition Variations across Titan's seas in a simple hydrological model, Geophysical Research Letters, 41, doi:10.1002/2014GL061133.

[12] Halley, E. (1714), A short account of the cause of the saltness of the ocean, and of the several lakes that emit no rivers; with a proposal, by help thereof, to discover the age of the world, Philos. Trans. R. Soc., 29, doi:10.1098/rstl.1714.0031

[13] A rather similar scenario was considered for Titan, where it was assumed that methane may steadily be converted by sunlight into ethane. Thus, by measuring the ethane:methane ratio in Titan's ocean (at the time it was assumed there was a single global ocean), then its age might be constrained.

[14] Halley, E. (1686). An Historical Account of the Trade Winds, and Monsoons, Observable in the Seas between and Near the Tropicks, with an Attempt to Assign the Phisical Cause of the Said Winds, By E. Halley. Philosophical Transactions, 16(179-191), 153-168.

[15] Mayer's work, from 1755, lay in some obscurity until published posthumously as Opera Inedita by Lichtenberg in 1771.

[16] Mairan made some early studies of the aurora borealis, as well as phosphorescent glows, but his most significant work was in biological clocks. He noted that certain heliotrope plants open and closed their leaves on a regular cycle even when in a dark room.

[17] G. B. Dalrymple, The Age of the Earth, 1994

[18] Bouger made a number of contributions in astronomy, and sailed to Peru on an expedition to measure the meridian arc. A measure of gravity variations, the Bouger anomaly, is named after him. Most of his later contributions were in fact in naval architecture.

[19] Franklin, B. (1755) Letter to Peter Collinson, dated Philadelphia, August 25, 1755, quoted in Lorenz, R. D., M. R. Balme, Z. Gu,, H. Kahanpää, M. Klose, M. Kurgansky, M. R. Patel, D. Reiss, A. P. Rossi, A. Spiga,T. Takemi, W. Wei. 2016. History and Applications of Dust Devil Studies, Space Science Reviews, 203, 5-37.

[20] Jefferson, T., Notes on the State of Virginia, 1787. A broader review of other outlooks of this time is J. R. Fleming, Historical Perspectives on Climate Change, Oxford, 1998

[21] As Fleming's book above notes, the role of trees in introducing moisture into the air was a noted effect at the time, of which even Christopher Columbus two centuries before had been aware. The possibility that land-clearing had changed the climate in Europe had been speculated upon (see e.g. K. Thompson, Forests and Climate Change in America: Some Early Views, Climatic Change, 3, 47-64, 1980), but the Americas offered the opportunity of a stronger experiment on this question.

[22] Herschel, W., 1801. Observations tending to investigate the nature of the Sun, in order to find the causes or symptoms of its variable emission of light and heat; with remarks on the use that may possibly be drawn from solar observations. Philosophical Transactions of the Royal Society of London, 91, pp.265-318.

[23] These ideas are discussed more fully in, for example, D. V. Hoyt and K. H. Schatten, The Role of the Sun in Climate Change, Oxford, 1997

[24] It is possible that Cornelis Drebbel may have replenished the oxygen in his submersible by cooking saltpetre

[25] In fact the Montgolfier brothers did not actually fly in the pioneering flight themselves, which was made by Pilatre de Rozier who later went on to develop a hybrid balloon (using hot air and light gas, combining advantages of both). Charles' gas balloon flight was viewed by some 400,000 spectators, among them was Benjamin Franklin, then the diplomatic representative to France of the new United States of America.

[26] An excellent biography of this fascinating man is Andrea Wulf's The invention of nature : the adventures of Alexander von Humboldt, the lost hero of science. London: John Murray, 2015. Humboldt's acquaintances read like a Who's Who of the period : considering just those mentioned elsewhere in the present boo, Humboldt stayed at the White House by Thomas Jefferson, gave financial support to Louis Agassiz, inspired Charles Darwin.

[27] Hamblyn, R.. The Invention of Clouds: How an Amateur Meteorologist Forged the Language of the Skies. Pan Macmillan, 2011.

[28] First and foremost, it is best to read Fourier's writings yourself, ideally in French. An English translation of his 1824 piece is found in E. Burgess, General Remarks on the Temperature of the Terrestrial Globe and the Planetary Spaces by Baron Fourier, American Journal of Science and Arts, 32, pp.1-20, 1837. Misquotations of Fourier are described in J. R. Fleming, Joseph Fourier, the 'greenhouse effect', and the quest for a universal theory of terrestrial temperatures, Endeavour, 23, 72-75, 1999. A very nice summary of early research on the greenhouse effect and glaciations, fittingly both in French and English, is E. Bard, Greenhouse effect and ice ages: historical perspective, Comptes Rendus Geoscience, 336, 603-638, 2004

[29] Well, mostly wrong. We are of course all immersed in a universal bath of about 3K of cosmic background radiation, and neither Earth's poles nor the other planets can get colder than this. So the picture is correct in a sense, but not right for explaining our polar temperatures.

[30] The term hydrometeors is still used in English as a catch-all for hailstones, raindrops, snowflakes etc., it is in this sense that Carnot's word 'meteor' is meant.

[31] Throughout this book I will unashamedly use both Centigrade (Celsius) degrees, and Kelvin as the occasion demands. Note that since the determination by the General Convention on Weights and Measures in 1967, the degree symbol ($^\circ$) is used with Centigrade, but not with Kelvin. Temperature (Kelvin)

= Temperature (Centigrade) + 273.15 K, thus room temperature is about 293 K.

[32] Smyth, C.P., 1856. Note on the constancy of solar radiation. *Monthly Notices of the Royal Astronomical Society*, *16*, p.220.

[33] In fact, De Saussure had noticed similar delay in propagation of the summer peak temperatures into the ground

[34] C. Piazzi-Smyth, Tenerife, An Astronomer's Experiment: or, Specialities of a Residence above the Clouds, Lovell Reeve, London 1858. The stereo pairs of photographs give striking 3-D viewing. On p.97 Piazzi-Smyth describes with some pride his construction of his wife's tent, with a built-in canvas floor that helped prevent it being blown away by wind, and also keeps out dust and sand. He also attempted, with a sensitive thermopile, to measure the heat of the moon, finding that the galvanometer needle twitched about the same from the moon as it did a candle 15 feet away. The book also discusses the effect of the reduced boiling temperature of water at these high elevations on the quality of tea.

[35] P.F.H. Baddeley, Whirlwinds and Dust Storms of India (Bell, London, 1860) See also Lorenz, R. D., M. R. Balme, Z. Gu,, H. Kahanpää, M. Klose, M. Kurgansky, M. R. Patel, D. Reiss, A. P. Rossi, A. Spiga,T. Takemi, W. Wei. 2016. History and Applications of Dust Devil Studies, Space Science Reviews, 203, 5-37

[36] A nice history of meteorology, and of weather maps in particular, is Mark Monmonier, Air Apparent – How Meteorologists Learned to Map, Predict and Dramatize Weather, University of Chicago Press, 1999

[37] the deflection of objects on a rotating sphere was recognized, in part, by Hadley, but there is an additional factor noted by Ferrel. Indeed, the deflection is often referred to as the Coriolis effect, after Gustave Coriolis but might be more properly attributed at least in part to Ferrel. A general relationship between pressure gradient and wind direction was noted by the Dutchman Buys Ballot in 1857.

[38] James Joule, who quantified the mechanical equivalent of heat and after whom the unit of energy is named, was a brewer. Hevelius, who mapped the lunar mountains, built his observatory with the profits from his brewery. The T-test, a widely-used statistical criterion, was published by William Gosset in the early 20th century under the pseudonym 'Student' because his employer,

Guinness of Dublin, had a policy against publication by its scientists. Science and beer have a long history together, even ignoring the contributions of Wilhelm Beer in mapping Mars.

[39] A very readable story of these early attempts by William Redfield, Elias Loomis and others, and the meteorological careers of Glaisher and Fitzroy, is Peter Moore, The Weather Experiment, Farrar, 2015

[40] Tyndall was born in Ireland which was at the time both areas were part of the United Kingdom ; Kelvin was born in Northern Ireland, which still is.

[41] The galvanometer was built by an acclaimed instrument-maker in Berlin, but this wasn't good enough, there was an annoying offset in the reading, which Tyndall traced to magnetic impurities in the wire. With some sleuthing, he traced that to some kind of dye in the silk insulation on the wire, which he replaced with white silk.

[42] J. Lequex (2009). Early Infrared Astronomy, Journal of Astronomical History and Heritage, 12(2), 125-140. Melloni used a Fresnel lens in part because it meant the overall thickness of the glass would be less than for a conventional lens, and glass absorbed infrared rays.

[43] Tyndall, J. (1861). The Bakerian Lecture: On the absorption and radiation of heat by gases and vapours, and on the physical connexion of radiation, absorption, and conduction. Philosophical Transactions of the Royal Society of London, 151, 1-36.

[44] Eunice Foote, Article XXL – Circumstances affecting the Heat of the Sun's Rays, read before the American Association, August 23, 1856. This crisp, straightforward paper appears as p.382-383 in Vol. XXII of The American Journal of Science and Arts, immediately after a rather muddled 5-page article by her husband Elisha Foote. Note that this neat observation is not a discovery of the greenhouse effect – if anything it is the opposite (i.e. an antigreenhouse effect, the absorption of sunlight rather than thermal infrared). The short-wave infrared absorption by carbon dioxide is very important, however, in the climate of Venus. Foote's observation is also hardly the earliest detection of near-infrared absorption – Fizeau and Foucault (of pendulum fame) recorded a solar spectrum in the near-infrared in 1847 that shows absorptions due to the Earth's atmosphere – see Lequeux note above.

[45] Ebelmen, J. J. (1847). Recherches sur la décomposition des roches. Carilian-Goeury et Vor. Dalmont. See also R. Berner, (2012). Jacques-Joseph Ébelmen,

the founder of earth system science. Comptes Rendus Geoscience, 344(11), 544-548.

[46] See M. E. Galvez and J. Gaillardet (2012), Historical Constraints on the Origins of the Carbon Cycle Concept, Comptes Rendus Geoscience, 344(11), 549-567

[47] J. Murray et al., Report on the Scientific Results of the Voyage of H.M.S. Challenger during the years 1873-1876, Her Majesty's Stationery Office, 1885 is still a fascinating volume. The Challenger's temperature soundings form an important benchmark on climate change – see D. Roemmich et al., 135 years of global ocean warming between the Challenger Expedition and the Argo Programme, Nature Climate Change, 2, 425-428, 2012.

[48] W. Leitch, God's Glory in the Heavens, Alexander Strahan, London 1867. The book imagines a journey through the solar system on a comet ('nature's rocket') and describes what might be seen. This appears to be one of the first mentions of rockets as devices for space travel, and of the superior performance in vacuum in particular. In 'Around the Moon', Jules Verne's 1870 sequel to 'From the Earth to the Moon', Verne imagines not only a barren moon seen by the protagonists through opera glasses from their bullet-space-ship, but also some small rockets intended to soften the bullet's landing being used for a course correction which sends it back to Earth.

[49] The superior performance of rockets in vacuum was not widely appreciated even 50 years after this – Robert Goddard's 1920 work was panned by the New York Times, which claimed incorrectly that rockets would not work in space.

[50] H. D. Taylor, (1895). Mars, A negative optical proof of the absence of seas in Mars. Monthly Notices of the Royal Astronomical Society, 55, 462-474.

[51] J. H. Poynting, Radiation in the Solar System: its Effect on Temperature and its Pressure on Small Bodies, Philosophical Transactions of the Royal Society of London, Series A. 202, pp.525-552. In fact, Arrhenius in his 1915 book attributes Christiansen with the first calculation, perhaps being more aware of work in Nordic countries than others.

[52] Christiansen was at the time the only Professor of Physics at the University of Copenhagen in Denmark. He made some advances in optics, and was the Ph.D. advisor of Niels Bohr (after whom the University is now named). Christiansen's original report "Nogle Bemaerkninger angaaende Planeternes Varmegrad" is in Danish, in the Oversigt over det Kong. Danske

Videnskarnernes Selskabs. Forhandlinger. Kjöbenhavn 1886 p. 85—108. but was summarized "Einige Bemerkungen über die Temperatur der Planeten" later that year in a German journal, Wiedemann's Beiblatter zu den Annalen der Physik und Chemie, Band X, 1886.

[53] Rosse's 6-foot telescope, The Leviathan, was the largest telescope in the world until the 100-inch Hooker telescope at Mt. Wilson in California in 1917. The 3-foot telescope was rather easier to use, which probably accounts for its application in this experiment.

[54] Earl of Rosse, On the Radiation of Heat from the Moon.--No. II , Proceedings of the Royal Society of London , Vol. 19 (1870 - 1871), pp. 9-14. His result is sometimes misquoted as saying the temperature of the moon reached 500F – that is not the same thing, since the night-time temperatures were high negative values on the Fahrenheit scale.

[55] Langley, S. P. The Temperature of the Moon, Science, Vol. 7 (158) pp. 8-9. A detailed account (170 pages) of Langley's experiments is 'The Temperature of the Moon', Third Memoir, National Academy of Sciences Vol. 4 part 2, printed in 1889. Langley's book, 'The New Astronomy' (1891) is a very readable description of the state of the art in the late 19th century.

[56] Langley, S. P. (1884). Researches on solar heat and its absorption by the earth's atmosphere: a report of the Mount Whitney expedition. US Government Printing Office.

[57] Arrhenius S. 1896. On the influence of carbonic acid in the air upon the temperature of the ground. The London,Edinburgh, and Dublin Philosophical Magazine and Journal of Science 41(251): 237–276

[58] T. Mellard Reade, Chemical Denudation in Relation to Geological Time, D. Bogue, London, 1879. Similar calculations were made by the Irish physicist John Joly in 1899 which seem to be better-known, but are also wrong. As Reade notes, the method can only establish a lower limit on the age, since salts can also be removed from the sea.

[59] Chamberlin, T. C. (1898). The influence of great epochs of limestone formation upon the constitution of the atmosphere. The Journal of Geology, 6(6), 609-621. See also Fleming, J. R. (2000). TC Chamberlin, climate change, and cosmogony. Studies in History and Philosophy of Science Part B: Studies in History and Philosophy of Modern Physics, 31(3), 293-308.

[60] Chamberlin, T.C., 1897. "A Group of Hypotheses Bearing on Climatic Changes." J. Geol. 5, 653–83.

[61] Chamberlin, T. C. 1906. On a possible reversal of deep-sea circulation and its influence on geological climates, J. Geology, 14, 363-373

[62] N. Ekholm, 1901. On the Variations of the Climate of the Geological and Historical Past and Their Causes, Quarterly Journal of the Royal Meteorological Society, 27, 11-62

[63] Now generally known as the Yule-Walker equations. Walker's mathematical interests were wide, and I had cause to cite his pioneering work on boomerang aerodynamics in my earlier book 'Spinning Flight', Springer, 2006. Udnay Yule was interested in exploring variations in sunspot numbers.

[64] Walker G. T. (1923). Correlation in seasonal variations of weather. VIII. A preliminary study of world-weather. Memoirs of the Indian Meteorological Department 24(Part 4) 75–131. The paper itself is rather heavy-going, being a discussion of various correlations. An illuminating review of Walker's contributions is Katz, R. W. (2002). Sir Gilbert Walker and a connection between El Nino and statistics. Statistical Science, 97-112. The formulation of ENSO as a combination of El Nino and Walker's Southern Oscillation was in fact made by Jacob Bjerknes, son of Wilhelm.

[65] Stoney, G. J. (1898). Of atmospheres upon planets and satellites. The Astrophysical Journal, 7, 25. In this paper he quotes his own work of some 31 years earlier, presented to the Royal Dublin Society, discussing how lighter gases should reach higher altitudes in planetary atmospheres, and that gas molecules moved fast enough to escape the Moon's gravity.

[66] A good summary is K. P. Hoinka, The tropopause: discovery, definition and demarcation, Meteorologische Zeitschrift, N.F. 6, 281-303, 1997.

[67] e.g. W. Dines, 1908. The Registering Balloon Ascents in England of July 22-27, 1907, Preliminary Account, Quarterly Journal of the Royal Meteorological Society, 34, 1-5. A. L. Rotch, 1897. On Obtaining Meterological Records in the Upper Air by Means of Kites and Balloons, Proceedings of the American Academy of Arts and Sciences, 32, 245-251.

[68] R. H. Goddard, A Method of Reaching Extreme Altitudes, Smithsonian, 1919

[69] The New York Times mocked these ideas at the time, claiming that a rocket had nothing to push against and so would not work in space. It thoughtfully published a retraction in 1969.

[70] The story is nicely told in W. Sheehan, The Planet Mars, A History of Observation and Discovery, University of Arizona Press, 1996

[71] I have astronomer colleagues who describe going outside the telescope dome at 5000m on Mauna Kea, site of many powerful telescopes because it is above much of the atmosphere, and report that very few stars are visible, because the thin air does not maintain low-light sensitivity in the eye. But when they take a whiff of pure oxygen from a mask (oxygen supplies are on hand in case of altitude sickness), they see an explosion of stars as their retinal function is temporarily restored. An elevation of about 2000m is roughly optimal between the improved clarity of the atmosphere at high altitudes and the improved sensitivity of the eye at lower altitudes.

[72] the observatory is now overlooked, bizzarely, by a rollercoaster in the Tibidabo amusement park further up the hill.

[73] R. D. Lorenz, 'Did Comas Sola discover Titan's Atmosphere?' Astronomy and Geophysics vol.38 No.3 June/July 1997, pp.16-18. In an odd coincidence, I actually went to kindergarten only a couple of km from this observatory. I had the opportunity to look through the telescope at a conference in Barcelona to celebrate the 5th anniversary of the Huygens probe descent.

[74] P. Lowell, A General method for evaluating the surface-temperature of the planets: with special reference to the temperature of Mars, The London, Edinburgh, and Dublin Philosophical Magazine and Journal of Science, Vol. 14 No.79, 161-176

[75] J. Poynting, On Prof. Lowell's Method for Evaluating the Surface-Temperatures of the Planets; with an Attempt to Represent the Effect of Day and Night on the Temperature of the Earth, The London, Edinburgh, and Dublin Philosophical Magazine and Journal of Science, 14(84), 749-760.

[76] A. R. Wallace, Is Mars Habitable, Macmillan, 1907. The book elaborates on Poynting's objections, as well as more general issues. Interestingly, it tells a fairly plausible story of planetary accretion, and suggests the canals are actually fractures formed by cooling of the planet, much like the cracks in basalt columns such as the Giant's Causeway.

[77] S. Arrhenius, The Destinies of Stars, Putnams, London 1918 (translated by J. Fries from the 1915 Swedish publication)

[78] An excellent dissection of the work in this period is C. Pekeris' D.Sc. Thesis 'The development and present status of the theory of the heat balance in the atmosphere', MIT, 1933, which can be found online. Some discussion is also found in R. Goody and Y. Yung's book, Atmospheric Radiation: Theoretical Basis, Oxford University Press, 1961.

[79] Emden, a Swiss working in Germany, in fact married Karl Swarzschild's sister.

[80] In the sun, light and heat are the same thing, so the meaning of gray atmosphere is clear. The Earth's atmosphere has quite different absorption of light and heat, as recognized since Fourier, and so two different shades of gray are assumed, one for heat and one for light, an approach usually termed 'semi-gray'.

[81] G. Simpson, 1927: Some Studies in Terrestrial Radiation. Memoirs of the Royal Meteorological Society 2(16) 69-95. He rather unimaginatively followed this up with G. Simpson, 1928: Further Studies in Terrestrial Radiation. Memoirs of the Royal Meteorological Society 3(21) 1-26. The next year's paper bravely broke the mould, G. Simpson, 1929: The Distribution of Terrestrial Radiation. Memoirs of the Royal Meteorological Society 3(23) 53-78.

[82] C. Abbe, 1901. The Physical Basis of Long-Range Weather Forecasts, Monthly Weather Review, 29, 551-561. Abbe needed consistent timekeeping to assemble weather reports, and in 1883 persuaded US railroad companies to adopt his system of four time zones across America. In 1903 Abbe wrote to Wilbur Wright to ask him to submit an article on wind currents to the Monthly Weather Review.

[83] Milutin Milankovic, 1879-1958, From his autobiography with comments by his son, Vasko and a preface by Andre Berger, European Geophysical Society, Kaltenburg-Lindau, 1995.

[84] M. Milankovich, Canon of Insolation and the Ice-Age Problem. The book – about 400 pages of algebra calculating sunlight variations - was published in Belgrade in 1941, just as Germany invaded Yugoslavia. The story of Milankovich, and the ultimate confirmation of the idea Adhemar, Croll and he

originated, is well-told in J. Imbrie and K. Imbrie, Ice Ages, Solving the Mystery, Harvard, 1979

[85] Richardson captured the essence of turbulence in his book with the famous verse 'Big whirls have little whirls that feed on their velocity, and little whirls have lesser whirls and so on to viscosity.' Richardson also discussed what would later be called 'fractals', finding that the measured length of a coastline depends on the length of the ruler one uses.

[86] E. Pettit and S. Nicholson, Radiation Measures on the Planet Mars, Publications of the Astronomical Society of the Pacific, 36, 269-272. Immediately following their paper is a report by Coblentz and Lampland, using similar methods but using a 40-inch telescope at Lowell. They claim the afternoon side was warmer than the morning, although don't assign numbers ; with 6 times less collecting area on their telescope, their observation was much more challenging and the results less certain. W. Sinton, Through the Infrared with Logbook and Lantern Slides, A history of Infrared Astronomy from 1868 to 1960, Publications of the Astronomical Society of the Pacific,98, 246-251, 1986 gives a concise review of these and related endeavours.

[87] The early work on Mars is nicely reviewed in D. Harland, Water and the Search for Life on Mars, Springer, 2005 and in W. Sheehan, The Planet Mars: A History of Observation and Discovery, U. Arizona Press, 1997

[88] Flower, W.D., 1936. Sand devils. Meteorological Office Professional Notes, 5, 1–16. HMSO London.

[89] Lorenz, R., 2013. The longevity and aspect ratio of dust devils: Effects on detection efficiencies and comparison of landed and orbital imaging at Mars. Icarus, 226(1), pp.964-970.

[90] Courant, R.; Friedrichs, K.; Lewy, H. (1928), Über die partiellen Differenzengleichungen der mathematischen Physik, Mathematische Annalen (in German) 100 (1): 32–74

[91] Bretz, J Harlen (1923), "The Channeled Scabland of the Columbia Plateau." Journal of Geology, v.31, p. 617-649. Soennichsen, John (2008), "Bretz's Flood: The Remarkable Story of a Rebel Geologist and the World's Greatest Flood", Seattle, Washington, Sasquatch Books

[92] V. R. Baker and D. Nummedal, 1978. The channeled scabland: a guide to the geomorphology of the Columbia Basin, Washington : prepared for the

Comparative Planetary Geology Field Conference held in the Columbia Basin, June 5-8, 1978 / sponsored by Planetary Geology Program, NASA Office of Space Science

[93] W. Koppen and A. Wegener, 1924. *Die Klimate der Geologischen Vorzeit.* Boerntrager. An english translation ("Climates of the Geological Past") was published in 2015. Milutin Milankovic apparently made significant contributions to the text, but is not named as an author. While usually associated with geology, Wegener wrote a book on the Thermodynamics of the Atmosphere in 1911.

[94] Wegener was a prominent Arctic explorer, and indeed himself perished on his skis on an expedition in Greenland

[95] G. S. Callendar. 1938. The artificial production of carbon dioxide and its influence on temperature. Quarterly Journal of the Royal Meteorological Society 64(275): 223–240. J. R. Fleming's The Callendar Effect, The Life and Work of Guy Stewart Callendar (1898-1964), American Meteorological Society, 2007 is an excellent story. It fails to notice, however, the strong undertone of dismissal in Simpson's discussion that followed Callendar's presentation of his work at the Royal Meteorological Society, recorded in the pages of the journal after the paper itself. Despite Simpson's eminence, Callendar's response to the comments on his paper was robust, and he went on in further work to improve his estimates in future papers, without changing the conclusions, which of course have been rather strikingly borne out.

[96] e.g. E. Hawkins and P. Jones, On Global Temperatures : 75 years after Callendar, Quarterly Journal of the Royal Meteorological Society

[97] Adams, W. S., & Dunham, T. (1932). Absorption bands in the infra-red spectrum of Venus. *Publications of the Astronomical Society of the Pacific*, 243-245.

[98] Wildt, R. (1940). Note on the Surface Temperature of Venus. The Astrophysical Journal, 91, 266-268. His greenhouse calculations were much less elaborate than those of Simpson, Callendar or others. He also made the claim that Venus had clouds of formaldehyde, a rather less successful idea than his greenhouse thoughts.

[99] The development of meteorology in the USA as a profession and as an academic discipline in the first half of the twentieth century, and the subsequent initial developments in numerical weather forecasting, are described in K.

Harper, Weather by the Numbers : The Genesis of Modern Meteorology, MIT Press, 2012

[100] Lewis, J. M. 2003. Ooishi's Observation Viewed in the Context of Jet Stream Discovery, Bulletin of the American Meteorological Society, 84, 357-369

[101] Ross Coen, Fu-go:The Curious History of Japan's Balloon Bomb Attack on America, University of Nebraska Press, 2014.

[102] A little-known British campaign was less meteorologically-ambitious, but more militarily successful. Operation Outward launched small balloons trailing wires from England towards Germany, where the wires shorted out power cables. Drapeau, Raoul E. 2011. "Operation Outward: Britain's World War II offensive balloons" . IEEE Power & Energy Magazine, 9, 94–105.

[103] Sverdrup, H.U., Munk, W.H., 1947. Wind, Sea and Swell: Theory of Relations for Forecasting. US Navy Hydrographic Office Publication No. 601, 44p.

[104] Lorenz, R.D. and Hayes, A.G., 2012. The growth of wind-waves in Titan's hydrocarbon seas. Icarus, 219(1), pp.468-475.

[105] And in fact with a relatively modest modern amateur astronomy setup, a small CCD camera with a diffraction grating, and an 8-inch telescope, you can repeat the discovery for yourself in about a minute of observing time. See Lorenz, R. D., Dooley, J. M., West, J. D., & Fujii, M. (2003). Backyard spectroscopy and photometry of Titan, Uranus and Neptune. Planetary and Space Science, 51(2), 113-125.

[106] H. Keiffer et al. The Planet Mars: From Antiquity to the Present, in H. Keiffer et al., Mars, University of Arizona Press, 1986. Another good historical account is D. M. Harland, Water and the Search for Life on Mars, Springer, 2005.

[107] Among WWII scientists working on radar were those working on systems for aircraft landing in fog, including Arthur C. Clarke (who in 1945 proposed that satellites in geostationary orbit would be useful telecommunications relays) and Luis Alvarez (who with his son Walter, would interpret an iridium-rich sediment layer at the Cretaceous-Tertiary geological boundary as evidence that an asteroid impact was responsible for the extermination of the dinosaurs.)

[108] Mayer, C. H., McCullough, T. P., & Sloanaker, R. M. (1958). Observations of Venus at 3.15-CM Wave Length. The Astrophysical Journal, 127, 1.

[109] Sinton, W. M., & Strong, J. (1960). Radiometric Observations of Venus. The Astrophysical Journal, 131, 470.

[110] Pettit, E., & Nicholson, S. B. (1955). Temperatures on the bright and dark sides of Venus. Publications of the Astronomical Society of the Pacific, 293-303. It is not obvious why it took them a quarter of a century to publish their results in full. Planetary astronomer Dale Cruikshank noted to me that Nicholson was primarily interested in stellar astronomy, which might explain the slow emergence of the work. Nicholson has the distinction, with Galileo, of having discovered four Jovian satellites.

[111] Of course, advanced computer developments were also underway in Britain for codebreaking ; these advances, however, remained under strict secrecy long after the war, and have only in the last couple of decades emerged (and have even become the stuff of movies), noting in particular the role of Alan Turing.

[112] Rossby was a prominent figure in both Scandinavian and American meteorological (and oceanographic) circles. He also established the journal Tellus, in which many important climatological papers would subsequently be published.

[113] Electronic Numerical Integrator and Computer – developed with gunnery applications in mind, this was at the US Army's Ballistics Research lab in Aberdeen, Maryland, a couple of hours drive from Princeton.

[114] P. Lynch, The Origins of Computer Weather Prediction and Climate Modeling, Journal of Computational Physics, 227, 3431-3444. This paper nicely sets these first experiments in context with Richardson and others. The scientific documentation of the first numerical experiment is Charney, J. G., Fjörtoft, R., & Von Neumann, J. (1950). Numerical integration of the barotropic vorticity equation. Tellus A, 2, 237-254.

[115] Phillips, N. A. (1956). The general circulation of the atmosphere: A numerical experiment. Quarterly Journal of the Royal Meteorological Society, 82(352), 123-164.

[116] Fultz, D. (1949). A preliminary report on experiments with thermally produced lateral mixing in a rotating hemispherical shell of liquid. Journal of Meteorology, 6(1), 17-33. Riehl, H., & Fultz, D. (1957). Jet stream and long

waves in a steady rotating-dishpan experiment: Structure of the circulation. *Quarterly Journal of the Royal Meteorological Society*, *83*(356), 215-231. Ghil, M., Read, P., & Smith, L. (2010). Geophysical flows as dynamical systems: the influence of Hide's experiments. Astronomy & Geophysics, 51(4), 4-28.

[117] H. Stommel, 1961. Thermohaline circulation with two stable regimes of flow, Tellus, 13, 224-241

[118] C. C. Langway, The History of Early Polar Ice Cores, US Army Cold Regions Research and Engineering Laboratory, ERDC/CRREL TR-08-1, January 2008

[119] Opik's rather wide-ranging career (both geographically as well as intellectually) is summarized briefly in C. Sterken, Ernst Julius Opik: Solar Variability and Climate Change, Baltic Astronomy, 20 195-203, 2011

[120] e.g. E. Opik, Ice Ages, Irish Astronomical Journal, 2, 71-84, 1952. Opik published even more obscure papers than this in the Communications of the University of Tartu – but his more easily obtained Icarus paper twelve years later (Opik, Climatic Change in Cosmic Perspective, Icarus, 4, 289-307, 1965) summarizes them.

[121] Wexler was the first scientist to deliberately fly in an aeroplane into a hurricane to collect scientific data in 1946; he was the chief scientist of the US Antarctic expedition in the IGY. Wexler later was a major advocate of satellites for weather observations, being a prime mover behind TIROS-1.

[122] Wexler, H. (1956). Variations in insolation, general circulation and climate. Tellus A, 8, 480-494 doesn't mention carbon dioxide at all. A chapter in H. Shapley, Climatic Change, Evidence Causes and Effects, Harvard University Press, 1953, does mention CO_2 in passing, but largely to dismiss it.

[123] e.g. Plass, G. N. (1956). The influence of the 15μ carbon-dioxide band on the atmospheric infra-red cooling rate. Quarterly Journal of the Royal Meteorological Society, 82(353), 310-324. and Plass, G. N. (1956). Effect of carbon dioxide variations on climate. American Journal of Physics, 24(5), 376-387. The contributions of Plass, Revelle, Suess and Keeling as well as Callendar, Tyndall, Arrhenius and others, are surveyed in Spencer Weart's (2008). The Discovery of Global Warming. Harvard University Press and many articles by that author.

[124] The bands are made up of lots of narrow lines. When viewed at low spectral resolution, the bands of CO_2 and H_2O overlap; but the overlap becomes less when looked at more closely as the individual lines that make the band do not overlap. The picture rapidly gets even more complicated, as the lines get broadened by the presence of even transparent gases like nitrogen, and start to overlap after all.

[125] Revelle, R., & Suess, H. E. (1957). Carbon Dioxide Exchange between Atmosphere and Ocean, and the Question of an Increase of Atmospheric CO2 during the Past Decades. Tellus, 9, 18-27

[126] Keeling, C. D. (1960). The concentration and isotopic abundances of carbon dioxide in the atmosphere. Tellus, 12(2), 200-203.

[127] Spinrad, H., Münch, G., & Kaplan, L. D. (1963). Letter to the Editor: the Detection of Water Vapor on Mars. The Astrophysical Journal, 137, 1319.

[128] P. Latil and T. Mar Planetary Observations by the multi-balloon technique, New Scientist, 7 May 1959, 1005-1007

[129] Charles Dollfus was the first Frenchman to cross the Atlantic in both directions by dirigible, and was instrumental in founding an aviation museum in Paris. He gave Adouin his first balloon ride at the age of 8.

[130] Dollfus, A. 1964. Observations of Water Vapor on Mars and Venus in The origin and evolution of atmospheres and oceans, Proceedings of a Conference, held at the Goddard Institute for Space Studies, NASA, New York, April 8-9, 1963. Edited by Peter J. Brancazio and A.G.W. Cameron. New York: Wiley, 1964., p.257. I met Dollfus at a conference dinner in 2004 in the Netherlands – even well into his 80s he was bright and engaging.

[131] House, F. B., Gruber, A., Hunt, G. E., & Mecherikunnel, A. T. (1986). History of satellite missions and measurements of the Earth radiation budget (1957–1984). Reviews of Geophysics, 24(2), 357-377.

INDEX

www.ingramcontent.com/pod-product-compliance
Lightning Source LLC
Chambersburg PA
CBHW021436170526
45164CB00001B/260